# Electromagnetism - Principles and Modern Applications:

## with Exercises and Solutions

## Essential Textbooks in Physics

Print ISSN: 2059-7630
Online ISSN: 2059-7649

The *Essential Textbooks in Physics* explores the most important topics in Physics that all Physical Sciences students need to know to pass their undergraduate exams (years 1, 2 and 3 of the BSc). Some topics are run-of-the-mill topics, others introduce students to more applied areas (e.g. Quantum Optics, Microfluidics…).

Written by senior academics as well lecturers recognised for their teaching skills, they offer in around 200 to 250 pages a theoretical overview of fundamental concepts backed by problems and worked solutions at the end of each chapter.

Their lively style, focused scope and pedagogical material make them ideal learning tools at a very affordable price.

Most authors are based at prestigious universities: Imperial College London, Oxford, UCL, Ecole Polytechnique.

*Published*

*Electromagnetism — Principles and Modern Applications: With Exercises and Solutions*
by Chris D. White

*Application-Driven Quantum and Statistical Physics: A Short Course for Future Scientists and Engineers*
*Volume 3: Transitions*
by Jean-Michel Gillet

*How to Derive a Formula*
*Volume 1: Basic Analytical Skills and Methods for Physical Scientists*
by Alexei A. Kornyshev and Dominic O'Lee

*Application-Driven Quantum and Statistical Physics: A Short Course for Future Scientists and Engineers*
*Volume 2: Equilibrium*
by Jean-Michel Gillet

More information on this series can also be found at https://www.worldscientific.com/series/etip

*(Continued at end of book)*

Essential Textbooks in Physics

# Electromagnetism - Principles and Modern Applications:
## with Exercises and Solutions

**Chris D. White**
*Queen Mary University of London, UK*

World Scientific

NEW JERSEY • LONDON • SINGAPORE • BEIJING • SHANGHAI • HONG KONG • TAIPEI • CHENNAI • TOKYO

*Published by*

World Scientific Publishing Europe Ltd.
57 Shelton Street, Covent Garden, London WC2H 9HE
*Head office:* 5 Toh Tuck Link, Singapore 596224
*USA office:* 27 Warren Street, Suite 401-402, Hackensack, NJ 07601

Library of Congress Control Number: 2022945020

**British Library Cataloguing-in-Publication Data**
A catalogue record for this book is available from the British Library.

**Essential Textbooks in Physics**
**ELECTROMAGNETISM — PRINCIPLES AND MODERN APPLICATIONS**
**With Exercises and Solutions**



ISBN 978-1-80061-361-4 (hardcover)
ISBN 978-1-80061-368-3 (paperback)
ISBN 978-1-80061-362-1 (ebook for institutions)
ISBN 978-1-80061-363-8 (ebook for individuals)

For any available supplementary material, please visit
https://www.worldscientific.com/worldscibooks/10.1142/Q0402#t=suppl

Desk Editor: Nur Syarfeena Binte Mohd Fauzi

Typeset by Stallion Press
Email: enquiries@stallionpress.com

For Michael and Toby.

# Preface

The origins of this book can be traced to my own experiences at the beginning of the 21st century. I was an undergraduate in physics, and struggling to get to grips with a theory that at high school had seemed perfectly innocuous. That theory was *electromagnetism* which, on a basic level, unifies the electric and magnetic phenomena we see around us into a single consistent description. It was first understood fully in the final years of the 19th century, and in our modern notation involves some rather advanced mathematical concepts, which are certainly not covered in a high school. However, this is only part of what makes learning electromagnetism daunting for the beginner. The final form of the theory is extremely abstract, and sufficiently so that one cannot sensibly start by writing down the final equations, and examining their consequences. Instead, it is common to follow a historical approach, in which one reconstructs (at least in broad brushstrokes) the key developments that led to our complete understanding. But following this approach too enthusiastically is incredibly inefficient, not least due to the fact that for most of the few thousand years that it took for electromagnetism to be tamed, our conceptual understanding of it as a collective species was either wrong or completely obscure. This is not helped by circumstance: some of the most familiar examples of electromagnetism we encounter, such as the bar magnets that we play with as children, are actually incredibly complicated systems from the point of view of theoretical physics. It is therefore sometimes simpler to bypass history, and to state the basic ideas of electromagnetism with a great deal of hindsight.

Another problem with beginning students of electromagnetism — particularly those as impatient as I was — is that they can often regard learning the theory as a chore to be endured, before reaching more "glamorous" theories later on. Traditional treatments of the subject do not always make clear the central role that electromagnetism plays in our understanding of all forces and matter in the universe, and how it arises in cutting-edge research in theoretical physics, including how to understand possible quantum theories of gravity. Students may encounter this through project work later in their studies, or at the beginning of a PhD programme, but there is no standard textbook that interpolates between the initial treatment of electromagnetism given to first year undergraduates, and the more advanced topics later on, much closer to the research frontier. The aim of this book is to redress this by offering, in a single and relatively brief text, an introduction to the complete theory of electromagnetism, together with some of its applications. This is then followed by chapters that explain the "covariant" formalism of electromagnetism that makes the requirements of Special Relativity manifest, and its connection with quantum field theory (QFT) — in particular how the theory can be obtained from an abstract mathematical symmetry principle called (*local*) *gauge invariance*. The latter material is commonly found in textbooks on QFT, but not in a form that is easily digested by undergraduates. The final chapter goes further, by showing how a highly active research topic in theoretical physics — that of finding relations between different (quantum) field theories — directly involves the ideas and methods of undergraduate electromagnetism. Throughout, I have aimed for an informal and conversational style, to make the more intimidating material seem slightly more accessible, and which emphasises ideas and their understanding over fussy technicalities.

This book serves a number of purposes for the modern physics student. It is suitable, for example, for a one or two-semester introductory course on electromagnetism, including some of its applications. It would then be useful later on in a physics degree, as a gentle introduction to advanced topics needed for understanding General Relativity and/or Quantum Field Theory. In bridging the gap between these less and more advanced subjects, it is natural that certain topics have had to be jettisoned in order to maintain a reasonable length. In particular, I have not explained in detail the behaviour of electric and magnetic fields in materials, apart from in very simple cases. Advanced aspects of electromagnetic waves, and the connection with (classical or quantum) optics, are also not discussed. However, a number of excellent books exist already on these subjects, and the reader of this book should be well-equipped to tackle them. On the more mathematical side, there are yet more formulations of electromagnetism that are not mentioned here. I do not, for example, use the advanced language of differential geometry (including differential forms and fibre bundles) in discussing either electromagnetism or gravity. This would take entire additional chapters to do justice to, for arguably little reward given the physical concepts already presented here in the more commonly used tensor language. I encourage the interested reader to find out more for themselves.

In summary, my hope is that this book both offers new ways of appreciating electromagnetism, and inspires readers to undertake further research themselves. I have written the book that I would have wanted in my own student days, and hope that it prevents others from suffering similar anxieties to my younger self!

A number of people have helped during the preparation of this book. I am first of all grateful to Laurent Chaminade at World Scientific for initiating the project, and to Nur Syarfeena Mohd Fauzi for detailed help as the proofs were being prepared. I owe a great debt to the students of Queen Mary University of London, who took my first-year electromagnetism course. Their passion and enthusiasm for physics were utterly infectious, and this book is my humble answer to their many interesting questions. I thank also my research collaborators Donal O'Connell and Ricardo Monteiro, whose work I have tried to do justice to in the final chapter. Finally, I thank my husband Michael and son Toby, to whom this book is dedicated, for their continued inspiration and support.

# Contents

# Chapter 1

# Why Electromagnetism?

The first thing to be done in any book is to motivate why it is worth reading it in the first place, so let me try to explain why we should all care about electromagnetism! To do this, I will first note that physics seeks to explain the world around us, and also to use the knowledge we gain to enhance our lives through technology. There are then, roughly speaking, two main areas:

(i) *Fundamental physics*: This aims to answer the big questions facing our existence, such as: How does the universe work? Why is it like it is? Where did it come from, and how will it end?
(ii) *Applied physics*: This is more concerned with industrial and technological aspects, and addresses such questions as: what are the practical applications of a theory? How does physics inform engineering, chemistry and biology? How do we make better technology?

The whole history of science teaches us that ideas from fundamental science have in turn become applied physics, and led to new technologies. This is not necessarily a quick process, but can take many decades. Examples include the theory of quantum mechanics, which first arose around 1900. Arguably the most important invention of the 20th century was the transistor, without which we would not have modern electronics or computers. Before the transistor, computers were the size of entire rooms, and had nothing like today's computing power! It was roughly 60 years between the beginnings of quantum theory (fundamental physics), and the commercial viability of working transistors (applied physics). A more striking example is the theory of General Relativity (1914), which revolutionised fundamental physics by predicting the unavoidable existence of curved spacetime, black holes and even the big bang itself. This is pretty exotic stuff that appears to have little in common with our everyday lives. However, nowadays General Relativity is also used to get your location accurate on your smartphone: your phone is sending signals to satellites in space, and the effects of spacetime curvature are important to get the calculations right!

These are just two examples of how fundamental and applied physics have interacted, but there remains a very fluid interface between these two areas, and it is important to cultivate both avenues of research. This doesn't just matter for the politicians who make decisions about how to fund science, but also for you personally as a budding scientist. As you read this, you may yourself be more inclined towards fundamental physics and the big questions, or towards more applied ends. However, your preference for different topics can change at different times (some of my favourite subjects now are ones that I absolutely hated as a student), and it pays to keep an open mind towards everything you encounter.

Now to return to *electromagnetism*. Although first fully understood in the late 1800s, observation of electrical and magnetic phenomena goes back thousands of years before that. However, do not let the long history of the subject fool you: despite being a relatively old theory, electromagnetism is still hugely important for fundamental physics. Our current understanding of nature is in terms of *matter*, which is acted on by *forces*. Furthermore, we currently believe that all of the forces we see around us are consequences of only four *fundamental forces*. The first two, electromagnetism and gravity, are familiar from everyday life. The additional two are called the weak and strong (nuclear) forces, and govern (amongst other things) the behaviour of atomic nuclei. Electromagnetism, together with the strong and weak forces, is described by the Standard Model of Particle Physics, which also includes the effects of quantum mechanics. Gravity, on the other hand, is described by General Relativity, and it remains an open research problem to unify this theory with quantum mechanics in a consistent way.

In the Standard Model, the weak and strong forces are described by generalisations of the equations describing electromagnetism, as we will see in Chapter 9. Thus, it really is true that you cannot understand the other forces in nature without first understanding electromagnetism itself. It is a simpler — but not necessarily a simple — theory, that we can often fall back on when trying to understand the more complicated forces of nature. We also have gravity to contend with, in the form of General Relativity. Traditionally, this was a very different theory to electromagnetism, and its equations look completely different. However, recent research around the world (since about 2008) suggests that GR could be much more closely related to electromagnetic-like theories than previously thought. We will see some of how this works in Chapter 10, but it is certainly true that electromagnetism is not a dead or useless subject, and that what you will learn in this book is much closer than you might think to the cutting edge of scientific research that is going on right now around the globe.

Electromagnetism was important for the history of fundamental physics in other ways too. We will arrive at the so-called *Maxwell equations* defining the theory, one of whose consequences is that they describe the nature of light. The speed of light in a vacuum ends up being constant for all observers, which ultimately led to the theory of Special Relativity. Furthermore, the Maxwell equations also anticipate quantum mechanics, as we will try to justify later on. Essentially, the equations have special abstract symmetry properties, that turn out to be needed for a quantum version of the theory to make sense. We will see these ideas in Chapter 9, and they apply to other forces in nature, such that they continue to inform our search for possible theories of quantum gravity.

So much for fundamental physics. For applied physics, electromagnetism is absolutely crucial, in a way that almost makes it too difficult to list. Think of how many technological devices you use in your everyday life for transport, communications, entertainment, grooming (of yourself or pets), work, study, etc. All of these involve electronics, and thus electromagnetism! Furthermore, you may yourself end up working in a branch of physics where you are involved in doing experiments (e.g. particle physics, astrophysics, cosmology, condensed matter physics, optics, etc.). Even if your immediate field of study does not appear to involve electromagnetism, it is essentially impossible to do any kind of measurement that does not involve turning something into an electric current so that it can be noticed and quantified.

If I have not yet convinced you of the importance of electromagnetism, let me add a political dimension. One of the biggest problems facing our globalised society is how less industrialised nations can develop alongside their more industrial counterparts, without exacerbating climate change. The same can be said for meeting the personal energy needs of everyone on the planet, without compromising the lifestyles we have come to enjoy. Many of the arguments in these areas of public policy involve global committees of politicians and world leaders. However, the solutions are going to have to be scientific/technological, and necessarily involve electromagnetism. Are you going to be one of the people that takes up or solves any of these grand challenges? I would like to think so, and if so you are reading the right book.

I hope I have managed to inspire some enthusiasm in you for learning electromagnetism. If so, you may need to try very hard to hold on to that enthusiasm as we proceed, as the subject we are about to learn can be intimidatingly complex. This is especially true given that we will need some very abstract mathematics that you may not have encountered before. But we will start on what I hope is relatively familiar territory. The main mathematical tools we will need throughout the book are *vectors* and *calculus*, which will end up being combined into the more general subject called, unsurprisingly, *vector calculus*. Let us start by reviewing some properties of vectors that you may well have seen before.

# Chapter 2

# Vector Algebra

Different quantities in nature can be represented by different types of mathematical object. Often, (real) numbers are sufficient for our purposes, e.g. for describing temperatures, speeds or masses. You will often see the word *scalar* to describe such quantities, and they are clearly not appropriate for everything. For example, a scalar is insufficient if a quantity has both a magnitude and a *direction*: if we are standing near a tiger and know it is travelling at 30 km $h^{-1}$, it would be useful to know if it is moving towards or away from us! For directed quantities, we can use *vectors*. In hand-written notes, these are usually written by placing an arrow over the quantity or underlining it. Thus, $\underline{v}$ and $\vec{v}$ both mean "vector $v$". In typeset documents, it is conventional to denote vectors in bold text, i.e. $\boldsymbol{v}$. We will adopt this convention throughout this book. The size of a vector is called the *magnitude*, or the *norm* (the latter term tends to be more common in more mathematical books). The symbol for this is to put vertical bars around the vector symbol so that, e.g. $|\boldsymbol{v}|$ means "the magnitude of vector $v$". We can then represent vectors by arrows, where the size of the arrow is equal to the magnitude of the vector, and the direction is simply the direction of the vector. Note that we shouldn't necessarily think of vectors as occupying a fixed location: in Figure 2.1, $\boldsymbol{a}$ is the *displacement* of $B$ relative to $A$ (i.e. the directed distance from $A$ to $B$); it is also the displacement of $D$ relative to $C$.

## 2.1 Addition and Subtraction

Given two vectors $\boldsymbol{a}$ and $\boldsymbol{b}$, we can add them as follows. If one places the head of vector $\boldsymbol{a}$ at the tail of vector $\boldsymbol{b}$, the vector $\boldsymbol{a} + \boldsymbol{b}$ is *defined* to be the vector that goes from the tail of $\boldsymbol{a}$ to the head of vector $\boldsymbol{b}$.

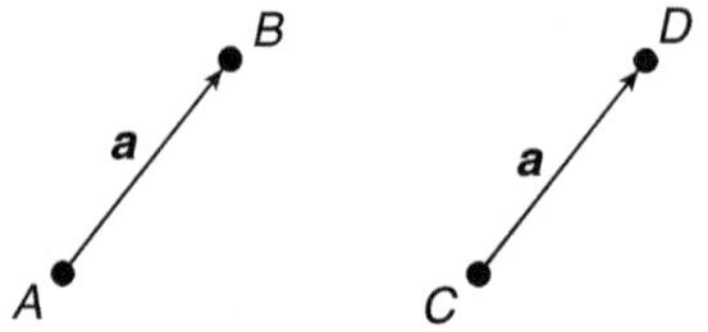


Fig. 2.1 The vector $\boldsymbol{a}$ represents the displacement of $B$ relative to $A$, but also of $D$ relative to $C$.

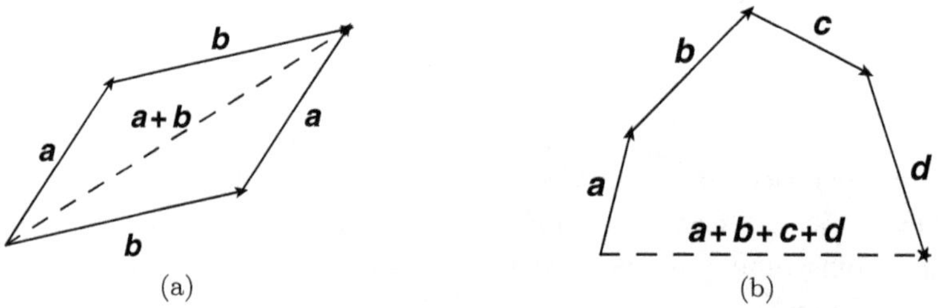


Fig. 2.2 (a) Definition of the vector $\boldsymbol{a}+\boldsymbol{b}$ (dashed arrow). The parallelogram shows that $\boldsymbol{a}+\boldsymbol{b}$ is the same as $\boldsymbol{b}+\boldsymbol{a}$; (b) addition of several vectors.

In other words, it forms the diagonal of the parallelogram spanned by $\boldsymbol{a}$ and $\boldsymbol{b}$. This is best illustrated by a diagram, such as Figure 2.2(a). The vector $\boldsymbol{a}+\boldsymbol{b}$ is sometimes referred to as the *resultant* of $\boldsymbol{a}$ and $\boldsymbol{b}$, and the addition procedure generalises easily to more than two vectors, as shown in Figure 2.2(b).

We can also note from Figure 2.2(a) that

$$\boldsymbol{a}+\boldsymbol{b}=\boldsymbol{b}+\boldsymbol{a}, \tag{2.1}$$

so that the order in which we add vectors is not important. The fancy way of saying this is that vector addition is *commutative*. One can also show that

$$(\boldsymbol{a}+\boldsymbol{b})+\boldsymbol{c}=\boldsymbol{a}+(\boldsymbol{b}+\boldsymbol{c}), \tag{2.2}$$

and the fancy way of describing this is that vector addition is *associative*.

Given vector addition, we can define vector subtraction by simply relabelling the diagram for addition. Let $\boldsymbol{c}=\boldsymbol{a}+\boldsymbol{b}$. Then Figure 2.2(a) can be redrawn as shown in Figure 2.3, where we show the vector obtained by subtracting $\boldsymbol{a}$ from $\boldsymbol{c}$ in red. We see that $\boldsymbol{c}-\boldsymbol{a}$ is a vector pointing from the head of $\boldsymbol{a}$ to the head of $\boldsymbol{c}$.

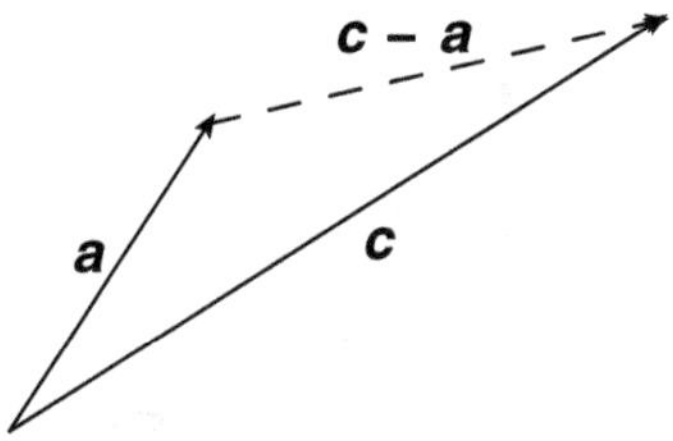


Fig. 2.3 Definition of vector subtraction, as obtained by relabelling the addition diagram of Figure 2.2(a). We see that $\boldsymbol{c} - \boldsymbol{a}$ is a vector pointing from $\boldsymbol{a}$ to $\boldsymbol{c}$.

After addition and subtraction, the next operation we can define on vectors is to multiply them by a scalar (number). For a real number $\alpha$ and a vector $\boldsymbol{a}$, we define $\alpha\boldsymbol{a}$ to be a vector with magnitude $\alpha|\boldsymbol{a}|$, pointing in the same direction as $\boldsymbol{a}$. That is, the size of $\alpha\boldsymbol{a}$ is $\alpha$ times the size of $\boldsymbol{a}$, but the direction is unchanged. If $\alpha$ is an integer, it is straightforward to verify that this definition is consistent with the previously defined rules for addition and subtraction. In general, you should always remember that multiplication of a vector by a scalar does not change the direction of the vector.

## 2.2 Vector Components

So far we have talked about vectors as abstract geometrical entities. In order to actually do calculations involving them, we need to be more specific, and the first thing to realise is that any vector $\boldsymbol{a}$ in three dimensions can be represented as a superposition of *basis vectors*. That is, let $\hat{\boldsymbol{i}}$, $\hat{\boldsymbol{j}}$ and $\hat{\boldsymbol{k}}$ be vectors with unit magnitude pointing in the $x$, $y$ and $z$ directions, respectively:

$$|\hat{\boldsymbol{i}}| = |\hat{\boldsymbol{j}}| = |\hat{\boldsymbol{k}}| = 1, \tag{2.3}$$

where, following convention, the hat symbol on a vector means that it has unit magnitude. Then a general vector $\boldsymbol{a}$ can be written as

$$\boldsymbol{a} = a_x\hat{\boldsymbol{i}} + a_y\hat{\boldsymbol{j}} + a_z\hat{\boldsymbol{k}}. \tag{2.4}$$

This procedure is illustrated in Figure 2.4, and the fact that the individual terms in Eq. (2.4) add up to give the total vector $\boldsymbol{a}$ follows directly from the addition rule for vectors that we defined in Section 2.1. As already mentioned above, the vectors $\hat{\boldsymbol{i}}$, $\hat{\boldsymbol{j}}$ and $\hat{\boldsymbol{k}}$ are referred to as (unit) basis

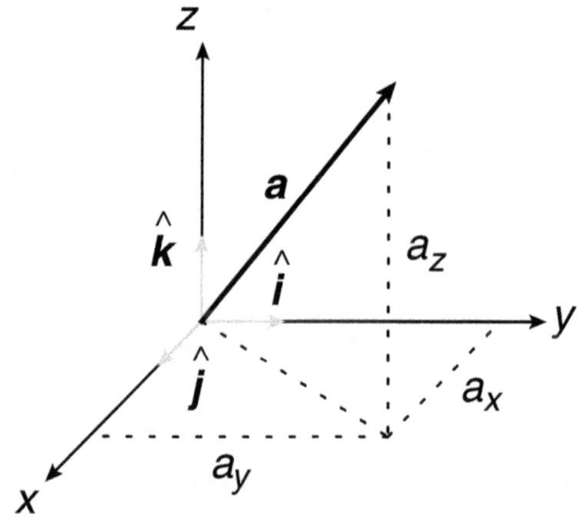


Fig. 2.4 A general vector $\boldsymbol{a}$ may be decomposed in terms of unit basis vectors $\hat{\boldsymbol{\imath}}$, $\hat{\boldsymbol{\jmath}}$, and $\hat{\boldsymbol{k}}$ (shown in light grey) parallel to the $x$, $y$ and $z$ axes respectively, where the geometric interpretation of the coordinates $(a_x, a_y, a_z)$ is shown.

vectors.[1] We could choose different basis vectors if we wanted to, but what matters is that there are enough basis vectors to fully describe any vector in the space: this is obviously true for the basis vectors we have chosen, as the $x$, $y$ and $z$ directions are all mutually orthogonal, and thus provide complementary information. We need three pieces of information to specify a vector in three dimensions, and thus three basis vectors.

The coefficients of the basis vectors in Eq. (2.4) are known as the *components* of $\boldsymbol{a}$, and they clearly depend on the choice of basis vectors. Were we to choose a different basis, the components of $\boldsymbol{a}$ would change, so that the vector on the left-hand side of Eq. (2.4) remains the same. For our choice of unit vectors in Cartesian coordinates, the geometric interpretation of the components is shown in Figure 2.4. Once we have fixed the basis, we often don't want to keep writing the basis vectors. Thus, the alternative notation

$$\boldsymbol{a} = \begin{pmatrix} a_x \\ a_y \\ a_z \end{pmatrix} \tag{2.5}$$

[1]In some books, you may see the notation $\hat{\mathbf{e}}_x$, $\hat{\mathbf{e}}_y$ and $\hat{\mathbf{e}}_z$ for the unit basis vectors in Cartesian coordinates.

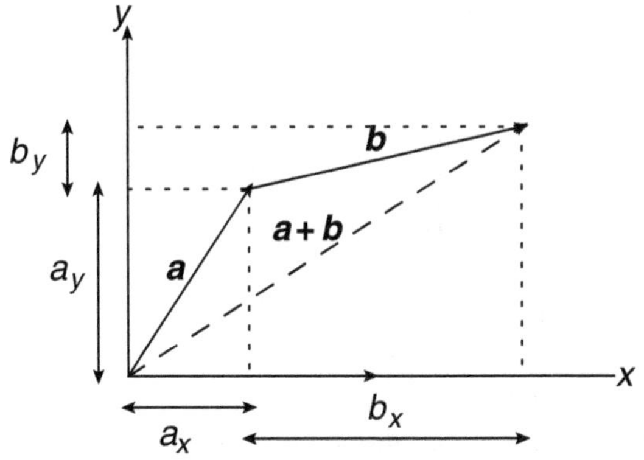


Fig. 2.5 Vector addition, with components labelled.

is used, in which the components are listed in a column. In this notation, vector addition takes the form

$$\boldsymbol{a}+\boldsymbol{b}=\begin{pmatrix}a_x\\a_y\\a_z\end{pmatrix}+\begin{pmatrix}b_x\\b_y\\b_z\end{pmatrix}=\begin{pmatrix}a_x+b_x\\a_y+b_y\\a_z+b_z\end{pmatrix}, \tag{2.6}$$

and an example is shown in Figure 2.5. By placing the vectors head to tail as dictated by the addition rule, the $x$ component of $\boldsymbol{a}+\boldsymbol{b}$ is indeed seen to be $a_x + b_x$, and likewise for any other components. Similar considerations apply for multiplication by a scalar, so that

$$\alpha\boldsymbol{a}=\alpha\begin{pmatrix}a_x\\a_y\\a_z\end{pmatrix}=\begin{pmatrix}\alpha a_x\\\alpha a_y\\\alpha a_z\end{pmatrix}. \tag{2.7}$$

We also need to know how to take the magnitude of vectors. From their geometric interpretation, this is simply given in terms of the components using Pythagoras' theorem[2]:

$$|\boldsymbol{a}|=\sqrt{a_x^2+a_y^2+a_z^2}. \tag{2.8}$$

We have here explicitly focused on three space dimensions, but for completeness it is worth noting that this result straightforwardly generalises to

[2]Although named after Pythagoras, the theorem was known from at least thousands of years before, possibly being independently discovered in various cultures. Written evidence survives from Mesopotamia, India and China.

any number $N$ of dimensions:

$$\text{if} \quad \boldsymbol{v} = \begin{pmatrix} v_1 \\ v_2 \\ v_3 \\ \vdots \\ v_N \end{pmatrix}, \quad |\boldsymbol{v}| = \sqrt{v_1^2 + v_2^2 + \cdots + v_N^2}.$$

Before leaving this section, it is worth mentioning that we have implicitly assumed a *right-handed* coordinate system throughout. Namely, upon rotating the $x$ axis into the $y$ axis with one's right-hand, the thumb denotes the direction of the $z$ axis. A different choice is possible, namely a *left-handed* coordinate system with an oppositely pointing $z$ axis. Physics cannot depend on which choice is made (this is just a human convention), but some of our formulae in what follows would change. We will follow convention by using a right-handed coordinate system throughout.

## 2.3 The Dot Product

As well as the operations defined in the previous section, it is useful to define certain more complicated operations on vectors, which are known as *vector products.* When first encountered, students can be confused as to where they come from, and the best answer I can give is that certain quantities involving vectors turn out to occur again and again throughout physics. Thus, it is useful to have a shorthand notation for these quantities, that simplifies the algebra in each case. If we did not define the products discussed below, we could still talk about the laws of electromagnetism, but they would look a lot messier!

The first product we will use is called the *dot product,* and is defined as follows:

$$\boldsymbol{a} \cdot \boldsymbol{b} = |\boldsymbol{a}||\boldsymbol{b}| \cos\theta, \tag{2.9}$$

where $\theta$ is the angle between the vectors $\boldsymbol{a}$ and $\boldsymbol{b}$, as shown in Figure 2.6.

From the Figure, we can recognise the *projection* of $\boldsymbol{a}$ onto $\boldsymbol{b}$, namely the component of $\boldsymbol{a}$ that is parallel to $\boldsymbol{b}$. This has magnitude

$$|\boldsymbol{a}| \cos\theta,$$

and thus one interpretation of the dot product is that it represents the projection of $\boldsymbol{a}$ onto $\boldsymbol{b}$, multiplied by the magnitude of $\boldsymbol{b}$.

From Eq. (2.9), the dot product is clearly a scalar, as the right-hand side contains only magnitudes of vectors, and an additional number. The

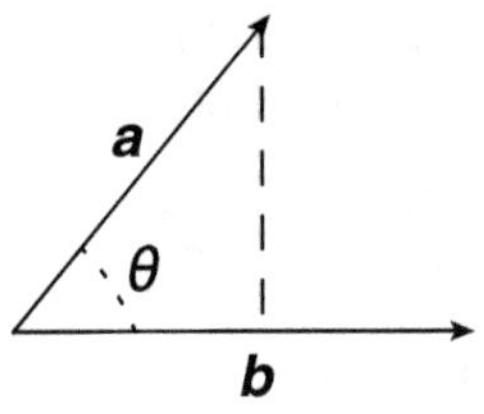


Fig. 2.6 The angle between two vectors $\boldsymbol{a}$ and $\boldsymbol{b}$, showing also the projection of $\boldsymbol{a}$ onto $\boldsymbol{b}$.

definition also implies that

$$\boldsymbol{a} \cdot \boldsymbol{b} = \boldsymbol{b} \cdot \boldsymbol{a} \tag{2.10}$$

(the dot product is commutative). Furthermore, one may show that

$$\boldsymbol{a} \cdot (\boldsymbol{b} + \boldsymbol{c}) = \boldsymbol{a} \cdot \boldsymbol{b} + \boldsymbol{a} \cdot \boldsymbol{c}, \tag{2.11}$$

which allows us to multiply out brackets. In mathematical language, Eq. (2.11) is referred to by saying that the dot product is *distributive.* Another consequence of Eq. (2.9) is that the dot product of two vectors vanishes if they are mutually perpendicular, which follows from the fact that $\cos(90°) = 0$. This in turn agrees with the above geometric interpretation: if two vectors are perpendicular, there is no projection of one onto the other. Another extreme example is taking the dot product of $\boldsymbol{a}$ with itself. Then $\theta = 0°$, and $\cos(0°) = 1$ implies

$$\boldsymbol{a} \cdot \boldsymbol{a} = |\boldsymbol{a}|^2, \tag{2.12}$$

which implies that the magnitude of any vector $\boldsymbol{a}$ can be found as $|\boldsymbol{a}| = \sqrt{\boldsymbol{a} \cdot \boldsymbol{a}}$, where the positive square root should be taken. It is useful to have a formula for the dot product in terms of the components of the two vectors. To this end, one may write

$$\boldsymbol{a} \cdot \boldsymbol{b} = (a_x\hat{\boldsymbol{i}} + a_y\hat{\boldsymbol{j}} + a_z\hat{\boldsymbol{k}}) \cdot (b_x\hat{\boldsymbol{i}} + b_y\hat{\boldsymbol{j}} + b_z\hat{\boldsymbol{k}}). \tag{2.13}$$

Next, we may use the distributive property of Eq. (2.11) to multiply out the brackets:

$$\begin{aligned} \boldsymbol{a} \cdot \boldsymbol{b} = {} & (a_x b_x \hat{\boldsymbol{i}}^2 + a_y b_y \hat{\boldsymbol{j}}^2 + a_z b_z \hat{\boldsymbol{k}}^2) \\ & + (a_x b_y + a_y b_x)\hat{\boldsymbol{i}} \cdot \hat{\boldsymbol{j}} + (a_x b_z + a_z b_x)\hat{\boldsymbol{i}} \cdot \hat{\boldsymbol{k}} + (a_y b_z + a_z b_y)\hat{\boldsymbol{j}} \cdot \hat{\boldsymbol{k}}, \end{aligned} \tag{2.14}$$

where we have used $\boldsymbol{a}^2$ as a shorthand for $\boldsymbol{a}\cdot\boldsymbol{a}$, for arbitrary vectors $\boldsymbol{a}$. On the first line, Eqs. (2.3) and (2.12) imply that

$$\hat{\boldsymbol{i}}^2 = \hat{\boldsymbol{j}}^2 = \hat{\boldsymbol{k}}^2 = 1.$$

On the second line, all the dot products vanish, as they all involve different basis vectors, which are mutually perpendicular. One finally obtains

$$\boldsymbol{a}\cdot\boldsymbol{b} = a_x b_x + a_y b_y + a_z b_z. \tag{2.15}$$

This is not too difficult to remember: to form the dot product of $\boldsymbol{a}$ and $\boldsymbol{b}$, you multiply the $x$, $y$ and $z$ coordinates separately, before adding them together. However, it is important to realise that Eq. (2.15) is only true in Cartesian coordinates. Were we to use a different coordinate system (e.g. polar coordinates), the results would be more complicated. Put another way, Eq. (2.15) is *not* the definition of the dot product. If we were shown it without knowing anything else, we would have no idea what it means! Equation (2.9) is instead the definition, that in principle holds in any coordinate system.

## 2.4 The Cross Product

It is also useful to have a second product of vectors, known as the *cross product*, and written as $\boldsymbol{a}\times\boldsymbol{b}$ or sometimes as $\boldsymbol{a}\wedge\boldsymbol{b}$. Unlike the dot product, this is a vector rather than a scalar, and the magnitude is defined via

$$|\boldsymbol{a}\times\boldsymbol{b}| = |\boldsymbol{a}||\boldsymbol{b}|\sin\theta. \tag{2.16}$$

The direction is defined to be perpendicular to $\boldsymbol{a}\times\boldsymbol{b}$, as given by the *right-hand rule*: upon curling the fingers of the right-hand from $\boldsymbol{a}$ into $\boldsymbol{b}$, the thumb points in the direction of $\boldsymbol{a}\times\boldsymbol{b}$ (if we had used a left-handed coordinate system, we would have used a left-hand rule here). This direction is illustrated in Figure 2.7. From the definition, we see that the cross product is not commutative, due to the fact that

$$\boldsymbol{a}\times\boldsymbol{b} = -\boldsymbol{b}\times\boldsymbol{a}. \tag{2.17}$$

That is, although the magnitude of the cross product does not care which order we multiply $\boldsymbol{a}$ and $\boldsymbol{b}$ in, the direction does, as can be verified using the right-hand rule — rotating $\boldsymbol{a}$ into $\boldsymbol{b}$ is the opposite of rotating $\boldsymbol{b}$ into

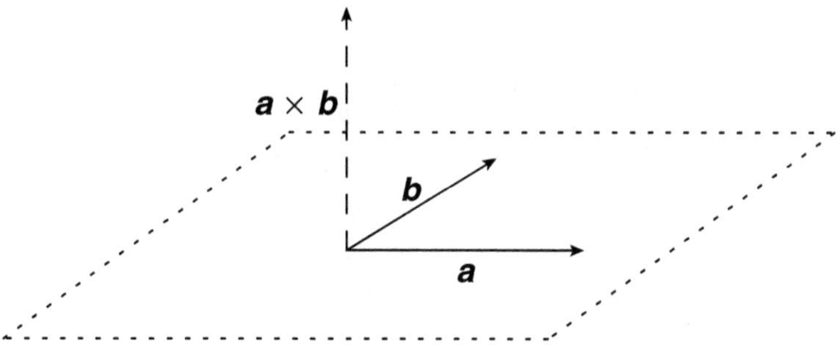


Fig. 2.7 The direction of the cross product $\boldsymbol{a} \times \boldsymbol{b}$ is perpendicular to the plane containing the two vectors $\boldsymbol{a}$ and $\boldsymbol{b}$, in a right-handed sense.

$\boldsymbol{a}$. The cross product is also not associative:

$$(\boldsymbol{a} \times \boldsymbol{b}) \times \boldsymbol{c} \neq \boldsymbol{a} \times (\boldsymbol{b} \times \boldsymbol{c}),$$

so beware! It does at least have the property of being distributive:

$$\boldsymbol{a} \times (\boldsymbol{b} + \boldsymbol{c}) = \boldsymbol{a} \times \boldsymbol{b} + \boldsymbol{a} \times \boldsymbol{c}, \tag{2.18}$$

which we can use to multiply out brackets involving the cross product.

Whereas the dot product involved the projection of $\boldsymbol{b}$ into $\boldsymbol{a}$, the magnitude of the cross product is instead sensitive to the component of $\boldsymbol{b}$ that is perpendicular to $\boldsymbol{a}$. From Eq. (2.16), we can note that the cross product will have zero magnitude if $\boldsymbol{a} \parallel \boldsymbol{b}$ (where $\parallel$ denotes "is parallel to"), given that $\theta = 0°$ for vectors which are parallel, and $\sin(0°) = 0$. We can get an explicit expression for the cross product of two vectors in terms of their components by first working out what the cross products of all the basis vectors are. It is a somewhat tedious exercise to show that the basis vectors satisfy

$$\begin{aligned} \hat{\boldsymbol{\imath}} \times \hat{\boldsymbol{\jmath}} &= -\hat{\boldsymbol{\jmath}} \times \hat{\boldsymbol{\imath}} = \hat{\boldsymbol{k}}; \\ \hat{\boldsymbol{\imath}} \times \hat{\boldsymbol{k}} &= -\hat{\boldsymbol{k}} \times \hat{\boldsymbol{\imath}} = -\hat{\boldsymbol{\jmath}}; \\ \hat{\boldsymbol{\jmath}} \times \hat{\boldsymbol{k}} &= -\hat{\boldsymbol{k}} \times \hat{\boldsymbol{\jmath}} = \hat{\boldsymbol{\imath}}. \end{aligned} \tag{2.19}$$

Then, using Eq. (2.18), the cross product of two vectors $\boldsymbol{a}$ and $\boldsymbol{b}$ is

$$\begin{aligned} \boldsymbol{a} \times \boldsymbol{b} &= (a_x\hat{\boldsymbol{\imath}} + a_y\hat{\boldsymbol{\jmath}} + a_z\hat{\boldsymbol{k}}) \times (b_x\hat{\boldsymbol{\imath}} + b_y\hat{\boldsymbol{\jmath}} + b_z\hat{\boldsymbol{k}}) \\ &= (a_yb_z - a_zb_y)\hat{\boldsymbol{\imath}} + (a_zb_x - a_xb_z)\hat{\boldsymbol{\jmath}} + (a_xb_y - b_ya_x)\hat{\boldsymbol{k}}. \end{aligned} \tag{2.20}$$

In column notation, this is

$$\begin{pmatrix} a_x \\ a_y \\ a_z \end{pmatrix} \times \begin{pmatrix} b_x \\ b_y \\ b_z \end{pmatrix} = \begin{pmatrix} a_y b_z - a_z b_y \\ a_z b_x - a_x b_z \\ a_x b_y - a_y b_x \end{pmatrix}, \tag{2.21}$$

which is not an immediately simple formula to remember. However, through sheer repetition and practice, many people tend to be able to remember it. Some people also use the following matrix determinant:

$$\boldsymbol{a} \times \boldsymbol{b} = \begin{vmatrix} \hat{\boldsymbol{i}} & \hat{\boldsymbol{j}} & \hat{\boldsymbol{k}} \\ a_x & a_y & a_z \\ b_x & b_y & b_z \end{vmatrix}. \tag{2.22}$$

This is only a mnemonic — it is a pretty strange matrix that has vectors and numbers in it as components! Furthermore, the trick is only useful if you know what a matrix determinant is, and can remember the formula for it. If you have not seen any such things before, then you can safely ignore this. As for the dot product, we should stress that Eq. (2.21) describes how to take the cross product in Cartesian coordinates only. Things would be considerably more complicated in other coordinate systems.

Having reviewed some properties of vectors, we are now in a position to start learning electromagnetism. The name suggests that we begin with the first part (electricity), and we shall indeed do this in the following chapter.

## Exercises

(1) Under what conditions is the dot product between two vectors zero? What about the cross product?

(2) Verify the conditions on the unit Cartesian basis vectors of Eqs. (2.19).

(3) Two vectors are given by

$$\boldsymbol{a} = \begin{pmatrix} 1 \\ 2 \\ 3 \end{pmatrix}, \quad \boldsymbol{b} = \begin{pmatrix} 3 \\ 2 \\ 1 \end{pmatrix}.$$

Calculate the angle between them from both the dot product and the cross product and show that you get the same answer.

(4) Does the cross product exist in two dimensions? What about more than three dimensions?

# Chapter 3

# Introducing Electricity

## 3.1 Electric Charge

Electric phenomena have been observed for thousands of years. Examples from the natural world include lightning (which can occur anywhere on the Earth's surface) and also in the animal kingdom. For instance, the *electric catfish* is native to tropical Africa and the Nile river. If encountering a scary human, it is liable to literally shock them, and it is interesting that at least two cultures identified the source of this shock with the same phenomenon (i.e. electricity) that causes lightning: the ancient Egyptians referred to the electric catfish as the "thunderer of the Nile"; similarly, 15th-century Arabs used the same word to refer to lightning as to a particular type of electric stingray. Other observations were made closer to modern-day Europe, e.g. the ancient Greeks noticed that rubbing materials such as amber with fur made them attractive or repulsive.

Our modern understanding of electricity is as follows. All everyday objects are made of atoms, which contain electrons orbiting nuclei made of protons and neutrons. Electrons and protons carry a fundamental property known as *charge*. This property has two types, which are given the labels *positive* and *negative*. These labels are arbitrary, and could easily have been called something else (or been named the other way around). The SI unit of charge is the *Coulomb*, with symbol C, and the charges of the electron and proton are $-e$ and $e$, respectively, where $e = 1.6 \times 10^{-19}$ C.

All observable particles in nature either have no charge, or charges which are integer multiples of $e$.[1] We thus say that charge is *quantised*— it is not a continuous variable, but exists only in discrete amounts. Another property of charge, that is verified in all experiments that have ever been carried out, is that it is *conserved*: in any given interaction, the sum of all charges in the initial state is equal to the sum of charges in the final state. This is even true when particles break up or decay into other particles, or when matter and antimatter are created in particle accelerators. Another way of saying this is that charge cannot be created or destroyed. It is perhaps natural to ask whether one can *derive* the existence of charge, or its conservation, from some more fundamental underlying principle. The answer to this is yes to a certain extent, and we will return to this in Chapter 9.

As stated above, whether or not charge is called positive or negative is merely a convention. What is physical, however, is that particles with the *same* type of charge repel each other, and particles with *different* types of charge attract each other. Furthermore, this force is incredibly strong when compared to, e.g. the force of gravity — roughly $2 \times 10^{39}$ times stronger in fact! There are many straightforward ways to demonstrate that electromagnetic forces are stronger than gravity. Think, for example, of rubbing a balloon to remove some of the electrons on it, creating a slight positive charge. This can be used to lift your hair, despite the fact that the mass *of the entire Earth* is trying to keep the hair down by the force of gravity! Given the huge difference between these forces, you may be wondering why it is gravity that we notice most of the time in our everyday lives, and not electromagnetism. This is because most objects around us have equal amounts of protons and electrons, to an extremely good approximation.

Different types of material exist, and the distribution of charge within a given material ultimately depends on how easy it is for charged particles to move around. If charges can move easily, the material is called a *conductor*; otherwise, it is called an *insulator*.[2] Solids can be conducting or insulating. They typically consist of fixed atomic centres comprised of nuclei and the inner electrons of an atom, plus the outer electrons of the atoms. The latter can either be strongly bound, in which case they can't move easily and the

[1] Strictly speaking, there are particles called (anti-)quarks that live inside protons and neutrons, as well as other types of composite particle. These can have charges $\pm\frac{e}{3}$ and $\pm\frac{2e}{3}$, but are always inside other particles so that they are not observed directly.

[2] There are other interesting possibilities (e.g. semiconductors, superconductors), but we won't worry about these in this book.

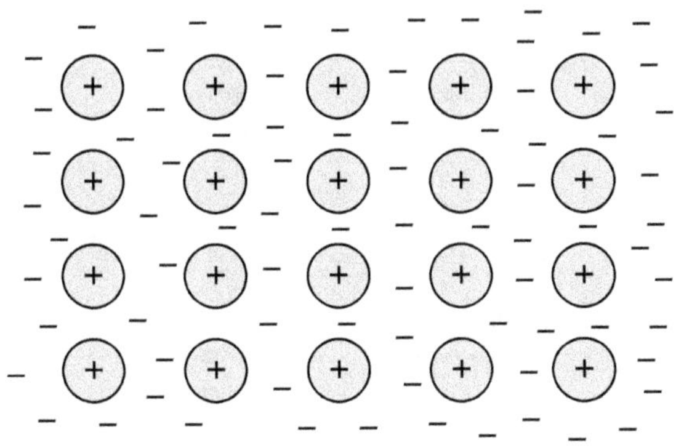

Fig. 3.1 Schematic view of a typical metal, consisting of relatively fixed ions, and a sea of weakly bound electrons (denoted by −) that are free to move throughout the material.

material is an insulator. Or the outer electrons can be weakly bound to the atomic centres, in which case they can move throughout the material and it is a conductor. This is represented highly schematically in Figure 3.1, which is a silly theorist's view of what goes on inside a metal. There is a lattice of positive ions, which are relatively heavy and thus immobile. Then there is a sea of weakly bound electrons distributed between the ions. It is these particles that move throughout the metal so that the material conducts electricity.

In fluids, particles are free to move in a way that is not possible in solids. However, fluids will still be insulating if they are made purely of electrically neutral particles (such as pure water, which contains neutral $H_2O$ molecules). Fluids become conducting if they contain charged particles (e.g. salty water, which contains $Na^+$ and $Cl^-$ ions). The situation is similar for gases, and a charged gas is usually called a *plasma*.

In general, charged particles can be either stationary or moving. The study of stationary charges is called *electrostatics*, and the study of moving charges is called *electrodynamics*. We will start with electrostatics, which is — perhaps unsurprisingly — simpler.

## 3.2 Coulomb's Law

To make progress in understanding electricity, we need to know the size of the force between two (static) charges, which can in principle be found out by doing lots of very careful experiments. However, at this point in history, the experiments have already been done, so we are allowed to simply quote

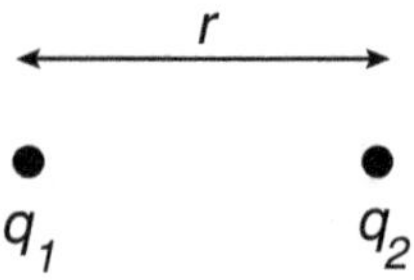


Fig. 3.2 Two charges $q_1$ and $q_2$ separated by a distance $r$.

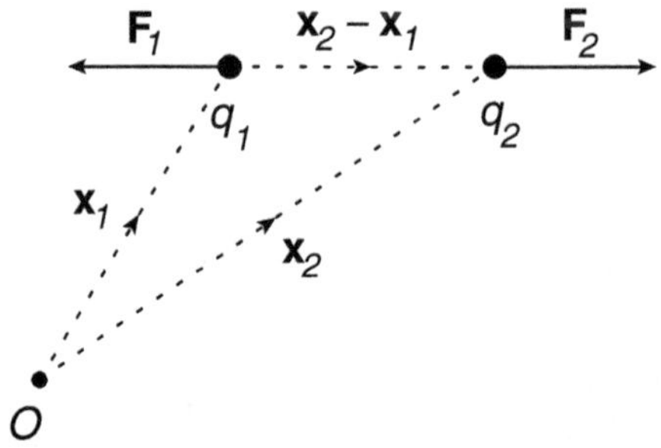


Fig. 3.3 The forces on two like charges, where $O$ is the origin, and $\boldsymbol{x}_i$ the vector position of the $i$th charge.

the answer! Let us take two stationary charges $q_1$ and $q_2$ separated by a distance $r$, as illustrated in Figure 3.2. Then the size of the force on $q_1$ due to $q_2$ is found to be

$$|\boldsymbol{F}| \propto \frac{q_1 q_2}{r^2}. \tag{3.1}$$

That is, it is proportional to the product of the charges, and inversely proportional to the square of the distance between them. For historical reasons, the constant of proportionality is usually written as $\frac{1}{4\pi\epsilon_0}$, so that the force is

$$|\boldsymbol{F}| = \frac{q_1 q_2}{4\pi\epsilon_0 r^2}. \tag{3.2}$$

This is usually referred to as *Coulomb's law*, and deserves some further comments. Firstly, the symmetry of Eq. (3.2) under interchanging $q_1$ and $q_2$ (i.e. the fact that the result stays the same) tells us that the size of the force on $q_2$ due to $q_1$ is the same as the size of the force on $q_1$ due to $q_2$ — an example of Newton's third law. Secondly, we have still not given the direction of the force. This is along the line joining the two charges, which we illustrate in Figure 3.3. Let us consider the case where the charges have the same sign, so that they repel each other. If $\boldsymbol{x}_i$ is the position of the

$i$th charge, then the vector $\boldsymbol{x}_2 - \boldsymbol{x}_1$ is a vector pointing from charge $q_1$ to charge $q_2$ (also shown in the figure). Thus, a unit vector in this direction is given by

$$\frac{\boldsymbol{x}_2 - \boldsymbol{x}_1}{|\boldsymbol{x}_2 - \boldsymbol{x}_1|}.$$

We can then write the vector force on charge $q_2$ due to charge $q_1$ as

$$\begin{aligned} \boldsymbol{F}_2 &= \frac{q_1 q_2}{4\pi\epsilon_0 r^2} \frac{\boldsymbol{x}_2 - \boldsymbol{x}_1}{|\boldsymbol{x}_2 - \boldsymbol{x}_1|} \\ &= \frac{q_1 q_2}{4\pi\epsilon_0} \frac{\boldsymbol{x}_2 - \boldsymbol{x}_1}{|\boldsymbol{x}_2 - \boldsymbol{x}_1|^3}, \end{aligned} \tag{3.3}$$

where we have used the fact that the separation of the charges $r$ is simply $|\boldsymbol{x}_2 - \boldsymbol{x}_1|$. Do not be misled by the cubic power in the denominator of the second line. The numerator contains a power of the distance, so that the overall force goes like one over distance *squared*, in accordance with Eq. (3.2). We can similarly write an expression for the force on charge $q_1$ due to charge $q_2$:

$$\boldsymbol{F}_1 = \frac{q_1 q_2}{4\pi\epsilon_0} \frac{\boldsymbol{x}_1 - \boldsymbol{x}_2}{|\boldsymbol{x}_1 - \boldsymbol{x}_2|^3} = -\boldsymbol{F}_2, \tag{3.4}$$

i.e. the forces point in opposite directions, also in accordance with Newton's third law. A further check of Eqs. (3.3) and (3.4) is that if $q_1 q_2 < 0$ (if the charges have opposite signs), the (vector) force flips sign, so that it becomes attractive rather than repulsive.

We have not yet given the constant $\epsilon_0$ a name. It is called the *permittivity of free space*, and its measured value in SI units is

$$\epsilon_0 = 8.85 \times 10^{-12}\,\text{N}^{-1}\text{m}^{-2}\text{C}^2. \tag{3.5}$$

As the name of $\epsilon_0$ suggests, Coulomb's law applies to charges in a *vacuum*. In a material, the force will be less, as charges can be *screened* by other charges. We will see examples of this later, but in simple cases the effect of the material is to modify $\epsilon_0 \rightarrow \epsilon_r \epsilon_0$, where $\epsilon_r > 1$ is called the *relative permittivity* of the material. It measures the degree to which a material permits electric forces to be produced, hence the name "permittivity"!

## 3.3 The Electric Field

Coulomb's law is a very powerful result, in that it allows us to calculate the forces on charges due to other charges. From Newton's laws, we can then work out how the charges will move, which goes a long way towards explaining electricity. However, there is something very philosophically weird about Coulomb's result — it seems to imply "action at a distance". Consider, for example, moving the charge $q_1$ in Figure 3.2 so that it is closer to the charge $q_2$. From Coulomb's law, the force on charge $q_2$ changes, which appears to happen instantaneously. However, this clearly contradicts Special Relativity, which says that information cannot travel faster than the speed of light. There must therefore be a delay in the information reaching $q_2$ from $q_1$, which also implies that there is something in between $q_1$ and $q_2$ through which the information can be propagated.

The modern viewpoint is that the charge $q_1$ sets up an *electric field*. Any other charge in an electric field feels a force, which is the origin of the Coulomb force felt by $q_2$. When we move the charge $q_1$, this creates a disturbance in the field, which then propagates to the charge $q_2$ no faster than the speed of light. It is equally valid to say that charge $q_2$ sets up an electric field, and that $q_1$ then feels a force due to the electric field set up by charge $q_2$. This idea generalises straightforwardly to arbitrary numbers of charges: any charge creates an electric field, where what we mean by this is that other charges feel a force, due to the presence of the field. Note that a given charge interacts with the electric field due to *other charges*, but not with its own electric field. If the latter were true, then objects could exert a net force upon themselves, which is impossible. The concept of fields was controversial when first proposed, but turned out to be extremely powerful. Indeed, the Maxwell equations that define electromagnetism are written in terms of the electric field (and the magnetic field, which we have yet to learn about). Nowadays, fields play a crucial role in our fundamental theories of nature, such as the Standard Model of Particle Physics. Field theory is also widely used throughout condensed matter physics, and in theories of the early universe (e.g. inflation).[3]

The problem of understanding electricity has now been shifted from forces, to understanding electric fields, and how they are related to forces. We can begin by more precisely defining the electric field as follows. Imagine placing a positive "test charge" $Q$ at a given point in space. Then the

[3] You may be wondering whether the above objection about "action at a distance" applies to Newton's law of gravity. Indeed it does, and the solution is the theory of General Relativity, as we discuss in Chapter 10.

electric field is defined as

$$\boldsymbol{E} = \frac{\boldsymbol{F}}{Q}, \tag{3.6}$$

where $\boldsymbol{F}$ is the force that the test charge experiences. Note that $\boldsymbol{E}$ is a vector, whose direction is the same as the net force experienced by a positive charge. This charge need not actually exist in practice: the electric field tells us about the force a positive charge *would experience*, were it placed in the electric field. Because we divide by the magnitude of the charge, the result does not depend on the strength of the charge that we place in the field. The SI units of the electric field can be written as $\mathrm{NC}^{-1}$ (Newtons per Coulomb), or alternatively as $\mathrm{kg\,m\,s^{-2}\,C^{-1}}$.

To give an example, imagine a charge $q_1$ at position $\boldsymbol{x}_1$, and a test charge $Q$ at position $\boldsymbol{x}_2$. From Eq. (3.3), the force experienced by the charge $Q$ is

$$\boldsymbol{F} = \frac{q_1 Q}{4\pi\epsilon_0} \frac{\boldsymbol{x}_2 - \boldsymbol{x}_1}{|\boldsymbol{x}_2 - \boldsymbol{x}_1|^3}.$$

Thus, from Eq. (3.6), the electric field at $\boldsymbol{x}_2$ is

$$\boldsymbol{E}(\boldsymbol{x}_2) = \frac{\boldsymbol{F}}{Q} = \frac{q_1}{4\pi\epsilon_0} \frac{\boldsymbol{x}_2 - \boldsymbol{x}_1}{|\boldsymbol{x}_2 - \boldsymbol{x}_1|^3}. \tag{3.7}$$

This is independent of the test charge $Q$, and only depends on the properties of the charge $q_1$, which is as it should be if we are to interpret this as the field generated by the charge $q_1$, that could influence arbitrary additional charges.

## 3.4 Principle of Superposition

Usually there is more than one charged particle in any given physical system. How then do we find the electric field due to multiple charges? The starting point is to note that the electric field is defined in terms of the force felt by a test charge. Then, recall that the *net force* on a particle is given by the (vector) sum of individual forces. We can thus apply this idea to work out how to combine the electric fields produced by multiple particles.

Consider a set of $N$ charges $\{q_i\}$ distributed throughout space, as shown in Figure 3.4. Now add a test charge $Q$. This will experience a force $Q\boldsymbol{E}_i$ due to each individual charge $q_i$, where $\boldsymbol{E}_i$ is the associated electric field,

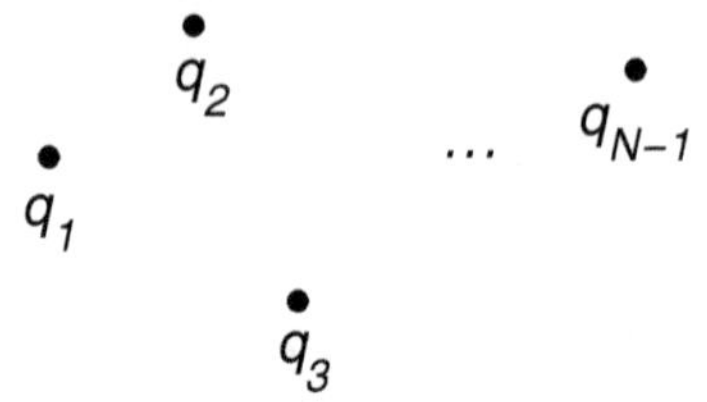


Fig. 3.4 A set of $N$ charges $\{q_i\}$ distributed throughout space.

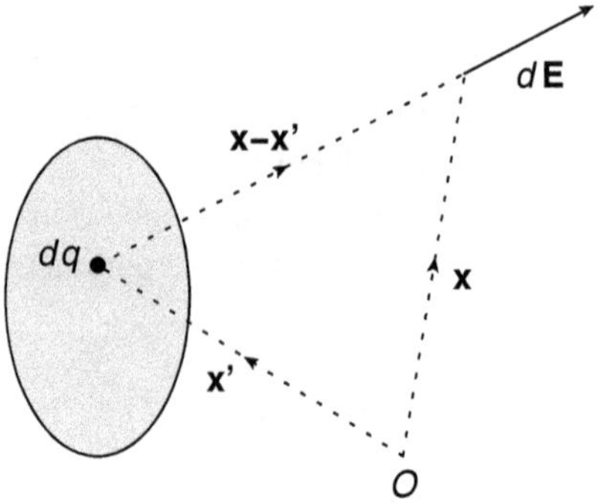


Fig. 3.5 A continuous distribution of charge, where the contribution to the electric field $d\boldsymbol{E}$ at $\boldsymbol{x}$ due to an infinitesimal portion of charge $dq$ at $\boldsymbol{x}'$ is labelled.

such that the net force on the charge $Q$ is

$$\boldsymbol{F} = \sum_{i=1}^{N} Q\boldsymbol{E}_i = Q\sum_{i=1}^{N} \boldsymbol{E}_i,$$

where we have used the fact that $Q$ is common to every term, and thus can be taken out of the sum. The total electric field due to all $N$ charges is given by Eq. (3.6), which implies

$$\boldsymbol{E} = \sum_{i=1}^{N} \boldsymbol{E}_i. \tag{3.8}$$

This is usually referred to as the *principle of superposition.* Namely, to find the electric field from a system of charges, one must superpose the electric fields due to individual charges (i.e. take the vector sum).

Having multiple point charges is still not the most general possibility. There can also be continuous distributions of charge, such as the shaded region in Figure 3.5. To find the field at a given point $\boldsymbol{x}$, we can split up the continuous distribution into infinitely many small portions of charge $dq$. In the figure, we consider the element of charge $dq$ at $\boldsymbol{x}'$. From Eq. (3.7), this contributes a field

$$d\boldsymbol{E} = \frac{dq}{4\pi\epsilon_0}\frac{\boldsymbol{x}-\boldsymbol{x}'}{|\boldsymbol{x}-\boldsymbol{x}'|^3}$$

at $\boldsymbol{x}$. Note that this field will be infinitesimal because $dq$ is, thus we label the field by $d\boldsymbol{E}$ accordingly. The total field can then be obtained by summing up all the fields due to the original charges $dq$. Due to the continuous nature of the charge distribution, this sum will be an integral, so that the total field is given by

$$\boldsymbol{E} = \frac{1}{4\pi\epsilon_0}\int \frac{\boldsymbol{x}-\boldsymbol{x}'}{|\boldsymbol{x}-\boldsymbol{x}'|^3}dq. \tag{3.9}$$

It is usually most convenient to describe a given continuous distribution by a *charge density* function $\rho(\boldsymbol{x}')$. That is, $\rho(\boldsymbol{x}')$ is the charge per unit volume at the point $\boldsymbol{x}'$. We can consider a small rectangular prism around the point $\boldsymbol{x}'$, such that the volume of this prism is given by

$$dV = dx'dy'dz',$$

where we have labelled coordinates according to $\boldsymbol{x}' = (x', y', z')$. If the sides of the prism are infinitesimally small, then the charge density $\rho(\boldsymbol{x}')$ is approximately constant inside it, so that the charge inside the prism is

$$dq = \rho(\boldsymbol{x}')dV = \rho(\boldsymbol{x}')dx'dy'dz' \equiv \rho(\boldsymbol{x}')d^3\boldsymbol{x}',$$

where the notation on the right is often used as a shorthand for $dx'dy'dz'$. This allows us to rewrite the integral for the electric field as an integral over all small prisms, namely the volume integral

$$\boldsymbol{E} = \frac{1}{4\pi\epsilon_0}\int\int\int \frac{\boldsymbol{x}-\boldsymbol{x}'}{|\boldsymbol{x}-\boldsymbol{x}'|^3}\rho(\boldsymbol{x}')d^3\boldsymbol{x}'. \tag{3.10}$$

In words, this means "multiply the charge density at each point in space by the infinitesimal volume of a prism around that point, and sum over all points in space". The three integral signs remind us that this is a 3D integral, as there is a single integral sign for each space dimension. If you have not seen the integral of a vector before, don't worry. All this tells us is that we can think of Eq. (3.10) as being three separate equations — one for each component of the electric field. One then integrates each component separately. Needless to say, volume integrals can be really quite complicated in

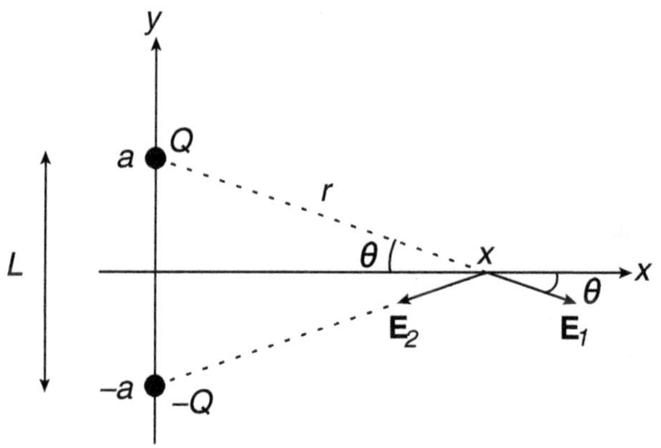


Fig. 3.6 A dipole positioned along the $y$-axis.

practice, although often symmetry can be used to simplify matters. Before moving on, let us give some examples of superposition in action!

(1) *Collection of N point charges*: For the system of point charges in Figure 3.4, the electric field at a general point $\boldsymbol{x}$ is given by combining Eqs. (3.7) and (3.8):

$$\boldsymbol{E} = \sum_{i=1}^{N} \frac{q_i(\boldsymbol{x} - \boldsymbol{x}_i)}{4\pi\epsilon_0 |\boldsymbol{x} - \boldsymbol{x}_i|^3}. \tag{3.11}$$

(2) *The electric dipole*: A *dipole* consists of a pair of opposite charges $\pm Q$ separated by a distance $L$. Figure 3.6 illustrates a dipole aligned with the $y$-axis, where the charges are placed at $y = \pm a$, $a = L/2$. The field due to a dipole at arbitrary points in space is rather complicated, but there are a couple of special cases that are tractable for all distances from the dipole:

(i) *Perpendicular to the dipole*: Consider a general point $x$ on the $x$-axis, as shown in Figure 3.6. We can work out the field at this point as the superposition of the fields $\boldsymbol{E}_1$ and $\boldsymbol{E}_2$ due to charges $Q$ and $-Q$, respectively. The first field is given by

$$\boldsymbol{E}_1 = \frac{Q}{4\pi\epsilon_0 r^2} \begin{pmatrix} \cos\theta \\ -\sin\theta \end{pmatrix},$$

where the column notation denotes the $x$ and $y$ components. This has the right magnitude using Coulomb's law and the definition of the electric field. The direction is obtained using trigonometry in Figure 3.6. Similarly, the second field is given by

$$\boldsymbol{E}_2 = \frac{Q}{4\pi\epsilon_0 r^2}\begin{pmatrix} -\cos\theta \\ -\sin\theta \end{pmatrix},$$

so that the superposition of fields gives a total field

$$\boldsymbol{E} = \boldsymbol{E}_1 + \boldsymbol{E}_2 = \frac{Q}{4\pi\epsilon_0 r^2}\begin{pmatrix} 0 \\ -2\sin\theta \end{pmatrix}. \tag{3.12}$$

This points in the $-y$ direction (downwards in the figure), with magnitude

$$|\boldsymbol{E}| = \frac{2Q\sin\theta}{4\pi\epsilon r^2}.$$

We can write this in terms of $x$ using Pythagoras' theorem

$$r = \sqrt{x^2 + a^2},$$

as well as

$$\sin\theta = \frac{a}{r} = \frac{a}{\sqrt{x^2 + a^2}}.$$

One thus finds

$$|\boldsymbol{E}| = \frac{2aQ}{4\pi\epsilon_0}\frac{1}{(x^2 + a^2)^{3/2}} = \frac{QL}{4\pi\epsilon_0}\frac{1}{(x^2 + a^2)^{3/2}}.$$

At large distances, this simplifies:

$$|\boldsymbol{E}| \xrightarrow{x \gg a} \frac{QL}{4\pi\epsilon_0}\frac{1}{x^3}.$$

(ii) *Along the line of the dipole*: Consider a point $y$ on the $y$-axis. The field due to charge $Q$ points in the $+y$ direction, and the field due to charge $-Q$ points in the $-y$ direction. One thus finds

$$\boldsymbol{E} = \frac{Q}{4\pi\epsilon_0}\begin{pmatrix} 0 \\ 1/(y-a)^2 \end{pmatrix} - \frac{Q}{4\pi\epsilon_0}\begin{pmatrix} 0 \\ 1/(y+a)^2 \end{pmatrix}.$$

This points in the $+y$ direction with magnitude

$$|\boldsymbol{E}| = \frac{Q}{4\pi\epsilon_0}\left[\frac{1}{(y-a)^2} - \frac{1}{(y+a)^2}\right] = \frac{QL}{2\pi\epsilon_0}\frac{y}{(y^2 - a^2)^2}.$$

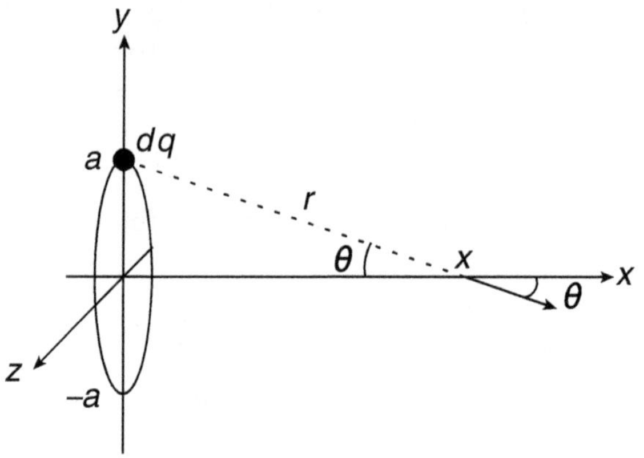


Fig. 3.7 A uniformly charged ring with radius $a$ and total charge $Q$, oriented in the $(y, z)$ plane.

At large distances this simplifies:

$$|\boldsymbol{E}| \xrightarrow{y \gg a} \frac{QL}{2\pi\epsilon_0} \frac{1}{y^3}.$$

We see that in either the $x$ or $y$ directions, the field of a dipole falls off faster than that of a point charge at large distances $r$. That is, it behaves as $r^{-3}$ rather than $r^{-2}$. This turns out to be quite general, for any direction.

(3) *A charged ring*: Consider a continuous, uniformly charged ring of radius $a$, with total charge $Q$, as shown in Figure 3.7. We can then ask: what is the field at a point $x$ along the $x$-axis (i.e. along the axis through the ring)? To answer this, we can split up the ring into infinitesimal segments, each with charge $dq$. One such segment is shown in the figure, and contributes to the $x$-component of the electric field at $x$ according to

$$dE_x = \frac{dq}{4\pi\epsilon_0 r^2} \cos\theta = \frac{dq}{4\pi\epsilon_0 r^2} \left(\frac{x}{r}\right) = \frac{dq}{4\pi\epsilon_0} \frac{x}{(x^2 + a^2)^{3/2}}.$$

The $y$ and $z$ components of the field due to this segment will also be non-zero. However, our aim is to find the total field, and the $y$ and $z$ components of the fields due to the all the segments combined will, by symmetry, cancel out. What I mean by this is that for every point on the ring, we can find a point on the opposite side of the ring with equal and opposite $y$ and $z$ components, so that there is no net field in these directions! Thus, the total field must be in the $x$-direction, and can be obtained by integrating the above expression over all charged segments:

$$\boldsymbol{E} = \hat{\boldsymbol{i}} \int \frac{dq}{4\pi\epsilon_0} \frac{x}{(x^2 + a^2)^{3/2}}.$$

Everything in the integrand is constant as far as the charge is concerned, i.e. the values of $x$ and $a$ are the same for all segments of the ring. Thus, we can replace the above with

$$\boldsymbol{E} = \hat{\boldsymbol{i}} \frac{1}{4\pi\epsilon_0} \frac{x}{(x^2 + a^2)^{3/2}} \int dq = \hat{\boldsymbol{i}} \frac{Q}{4\pi\epsilon_0} \frac{x}{(x^2 + a^2)^{3/2}},$$

where $Q = \int dq$ is (by definition) the total charge. Note that at large distances, one has

$$\boldsymbol{E} \xrightarrow{x \gg a} \frac{Q}{4\pi\epsilon_0 r^2},$$

so that the ring looks like a point charge. This makes sense given that if we zoom out, we can't resolve the structure of the ring, and it must look point-like. This was not true for the dipole, as that was electrically neutral overall (i.e. it contained opposite charges $\pm Q$).

## 3.5 Field Lines

One way to visualise the electric field is to draw arrows at each point in space, representing the magnitude and direction of the $\boldsymbol{E}$ vector. Another useful way to visualise them is to use *field lines.* These begin on positive charges and end on negative charges, such that the direction of the field line is the direction of the electric field (that is, $\boldsymbol{E}$ is always tangent to a field line). Two examples are shown in Figure 3.8. The first shows a positive point charge, for which we know the electric field always points in the radial direction at any point in space. The field lines are thus radial lines, and the arrow indicates the direction of $\boldsymbol{E}$ along the line. If we chose a negative point charge, the arrows would be pointing inwards. The second example in Figure 3.8 is that of an electric dipole. Above, we saw that the field should be in the vertical direction perpendicular to the centre of the dipole, or along the axis. This can indeed be seen to be the case. The direction of the field at other points is more complicated, as the field lines branch out from the positive charge, and loop round before ending on the negative charge.

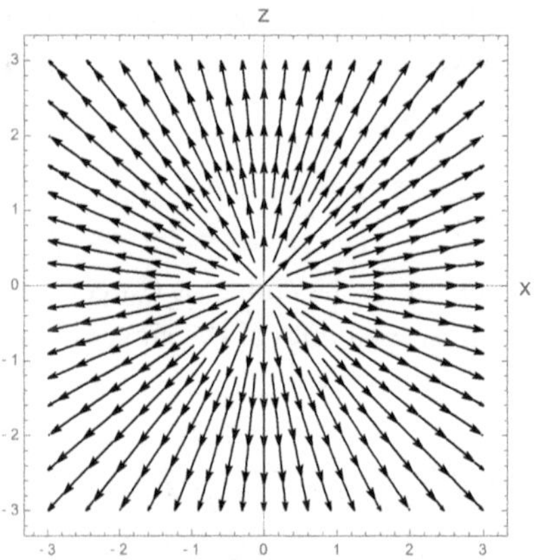


(a)

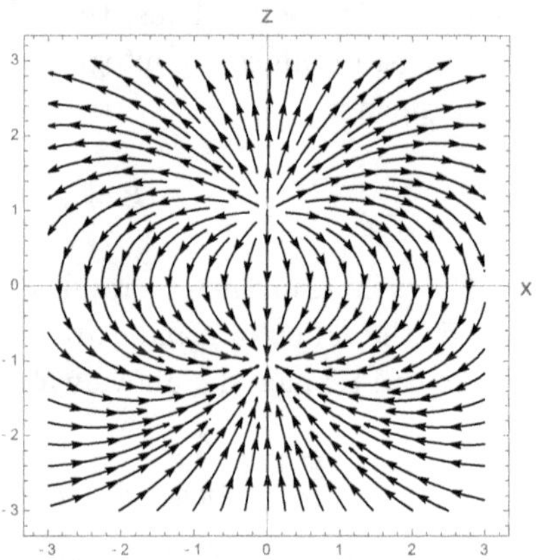


(b)

Fig. 3.8 The field lines due to (a) a positive point charge located at the origin; (b) an electric dipole, with positive and negative charges placed at $(x, y, z) = (0, 0, 1)$ and $(0, 0, -1)$, respectively.

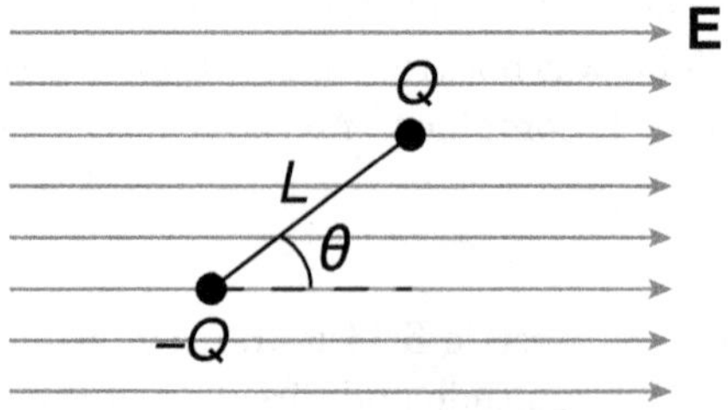


Fig. 3.9 A dipole of length $L$ placed at angle $\theta$ to a uniform applied electric field $\boldsymbol{E}$.

Note that field lines do not overlap with each other, or intersect. If this were the case, the electric field at the point of intersection would be ambiguous, which is physically nonsensical. Furthermore, for a given set of field lines, the lines are closer together when the field is strong, and further apart when the field is weak (e.g. for the point charge, we know the field falls off to zero at infinity due to being $\propto r^{-2}$). Given that $\boldsymbol{E}$ is tangent to a field line, a given line tells us which direction a positive charge would start to move in if we placed it in the field. A negative charge would move in the opposite direction.

As an example, consider a dipole in a uniform applied electric field, as shown in Figure 3.9. From the field lines, we can see that a positive charge feels a force to the right, whereas the negative charge feels a force to the left. The magnitude of both of these forces is $QE$, where $E \equiv |\boldsymbol{E}|$. This creates a *torque* $\boldsymbol{G}$ about the centre of the dipole, that tries to rotate it to align with the applied field. Given that the torques associated with both forces are in the same rotational direction (clockwise), the total magnitude of the torque is simply twice the torque associated with one of the forces. Thus, one has

$$|\boldsymbol{G}| = 2 \times (F \sin\theta) \times \frac{L}{2},$$

where $F \sin\theta$ is the size of the component of force perpendicular to the dipole, and $L/2$ the distance from the charge $Q$ to the pivot point (the centre of the dipole). Given $F = QE$, this gives

$$|\boldsymbol{G}| = QLE \sin\theta. \tag{3.13}$$

The direction of the torque vector is given by the right-hand rule: if the fingers of the right hand curl in the direction of the rotation, the thumb gives the direction of the torque vector. Thus, the torque vector is pointing into the page.

The quantity $QL$ occurs very often when talking about dipoles. Thus, it is conventional to define the *dipole moment* $\boldsymbol{\mu}$, which is a vector pointing from the negative charge to the positive charge, with magnitude $|\boldsymbol{\mu}| = QL$ (i.e. this is the product of the charge at one end of the dipole, and the length of the dipole). Then we can write an exact vector equation for the torque on a dipole:

$$\boldsymbol{G} = \boldsymbol{\mu} \times \boldsymbol{E}. \tag{3.14}$$

The magnitude of the cross-product can be seen to reproduce Eq. (3.13) above. You can check from the right-hand rule for cross products that the direction of Eq. (3.14) is indeed into the page.

We have mentioned dipoles twice now, and you may be wondering why. The answer is that dipoles occur many times in various branches of physics. As an example, consider diatomic molecules (small molecules containing two atoms only). If these atoms are different elements, there will be a charge imbalance in the molecule, so that each end is either positively or negatively charged. However, the total charge of the molecule will be zero, so that it behaves just like an electric dipole! We will revisit this later, when we look at what electric fields do to materials. Chemists also use these facts when they perform *rotational spectroscopy*, which is a fancy term for sticking some molecules in an electric field, measuring how they rotate, and using the results to find out what the molecules are. Another widespread use of electric dipoles is in communications: it turns out that shaking a dipole leads to electromagnetic radiation being emitted. By controlling how it is shaken, you can code the radiation with messages that can then be received by someone else. This is more commonly called an "antenna", and you are doing this every time you phone or text someone.

## 3.6 Electric Flux

Having introduced the electric field $\boldsymbol{E}$, our next aim is to understand the equations which govern its generation and evolution. It is an example of a *vector field*, meaning that it associates a vector with every point in space-time. Given a vector field, one can define the concept of *flux*, which turns out to be very useful, as we will see. To introduce the idea of flux, it is useful to consider another example of a vector field that is more familiar from everyday life, namely the flow of a fluid. At each point in space, the fluid at that point will have some velocity. Thus, the flow of the fluid can be represented by a field of velocity vectors. The simplest case is if the flow is uniform, so that the velocity of the field is the same at each point. This situation is shown in Figure 3.10(a), which also shows a surface of area $A$ constructed perpendicular to the fluid. In a given time period $\Delta t$, any fluid

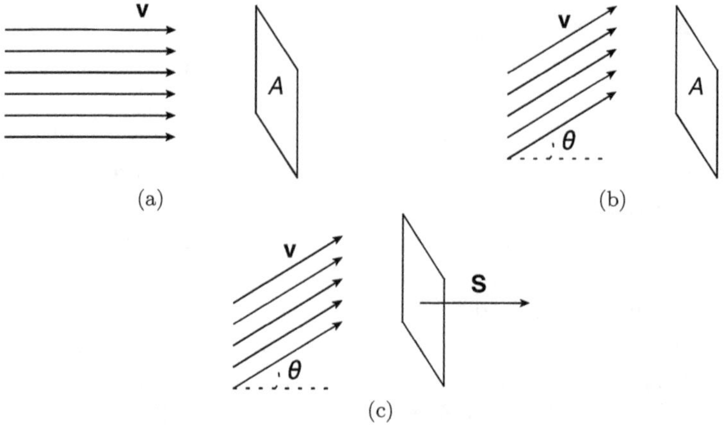


Fig. 3.10 (a) A uniform fluid flow incident on a perpendicular area $A$; (b) a uniform fluid flow directed at an angle $\theta$ with respect to an area $A$; (c) as for (b), but with the vector area $\boldsymbol{S}$ depicted.

that is less than a distance $|\boldsymbol{v}|\Delta t$ to the left of the surface will be able to cross it (any further than this, and it will not make it in time). Thus, the volume of fluid that crosses the surface is given by

$$|\boldsymbol{v}|A\Delta t,$$

so that the volume crossing per unit time is

$$|\boldsymbol{v}|A,$$

which represents the "rate of flow of fluid". Now consider a velocity field which is uniform, but not perpendicular to the area $A$. The amount of fluid crossing the surface in time $\Delta t$ is then

$$|\boldsymbol{v}|\cos\theta A,$$

as it is only the component of velocity perpendicular to the area that leads to fluid crossing the surface. We can write this in a more convenient form by defining a *vector area* $\boldsymbol{S}$, whose magnitude is $A$, and which points outwards from the surface, as shown in Figure 3.10(c). Then the rate of flow of fluid across the surface is

$$\Phi = \boldsymbol{v} \cdot \boldsymbol{S}.$$

This is known as the *flux* and, as we have seen, it represents the rate of flow of fluid across a surface.

The electric field is much more abstract to think about than a flowing fluid. However, if we want to, we can still define the concept of a "flux" for *any* vector field. We do indeed want to do this, as it turns out to be a very key concept needed to write down the equations of electromagnetism! Try not to worry if it seems too abstract at first: all we mean by the flux is some sort of measure of how much of the vector field is crossing a given surface in space. We do not necessarily have to try and interpret this as the flow of some weird "fluid", but having pictures in our minds involving arrows crossing a surface allows us to visualise the flux in a useful way. After all, trying to form mental pictures of what's going on, however silly they are, is one way that physicists cope with complicated maths.

Returning to the electric field, let us consider a given plane surface with vector area $\boldsymbol{S}$, and define the *electric flux*

$$\Phi_E = \boldsymbol{E} \cdot \boldsymbol{S}. \tag{3.15}$$

Above, we have restricted to the special case of plane surfaces, which have the property that the vector area is constant as a function of position. We

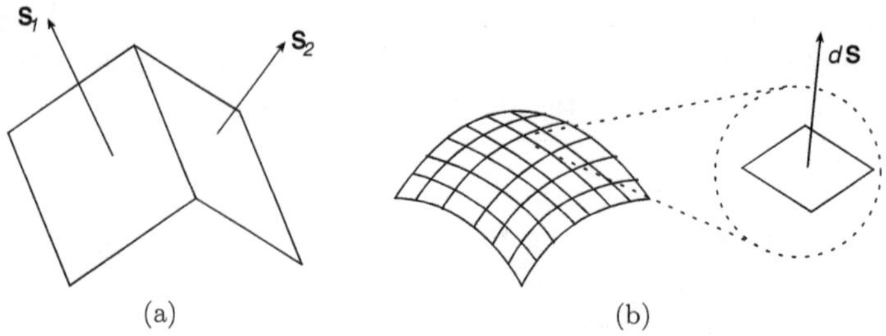


Fig. 3.11 (a) A surface composed of two separate planes, with vector areas $\boldsymbol{S}_1$ and $\boldsymbol{S}_2$; (b) A general curved surface looks locally flat, such that one can define the flux through each infinitesimal segment.

also considered the case of uniform electric fields. In general, however, we can define the flux for any arbitrary curved surface, and for electric fields which are spatially varying. Consider first the case of two plane surfaces joined together, as shown in Figure 3.11(a). Any stuff flowing through the surface must go through one part of the surface or the other. Thus, the total flux must be a sum of the fluxes through the two separate parts. Carrying this idea further, a general curved surface is shown in Figure 3.11(b). Provided the surface is sufficiently smooth, we can zoom in to a point on the surface, around which the surface will look flat (think for example about how, on the surface of the Earth, things look locally flat, even though the Earth's surface is curved). We can then divide the surface into infinitesimally small segments, each of which has some area $dS \equiv |\boldsymbol{dS}|$, where $d\boldsymbol{S}$ is the vector area pointing outwards from the segment (see Figure 3.11(b)). For an infinitesimal surface element, the electric field will be approximately constant on it, and thus the flux through the element at position $\boldsymbol{x}$ will be given by

$$d\Phi_E = \boldsymbol{E}(\boldsymbol{x}) \cdot d\boldsymbol{S}(\boldsymbol{x}),$$

where we have made clear that both the electric field and the vector area can depend upon position. The total flux is then obtained by summing over all the surface elements. Due to the continuous nature of the surface, this is an integral, and is usually written as

$$\Phi_E = \int\int_S d\boldsymbol{S} \cdot \boldsymbol{E}. \tag{3.16}$$

The two integral signs remind us that this is a double integral (i.e. over two parameters defining a surface). One reads the symbol $\int\int_S$ as "the integral over the surface $S$", and such integrals are called *surface integrals* for short. As for the volume integrals we discussed previously, surface integrals can be very complicated in general, and we discuss this full complication in appendix A. However, for all of the situations we will actually encounter in this book, the surface integrals will simplify, due to the amount of symmetry involved in each situation. As an example, consider an electric field that points radially outwards from the origin, and whose magnitude depends only on the radial distance $r$ from the origin. Such a field has *spherical symmetry*, meaning that all directions in 3D space look the same. Now consider a spherical surface $S$ of radius $R$. The vector area at each point on the sphere will point radially outwards, as shown in Figure 3.12(a). Thus, for all infinitesimal area segments on the sphere one has

$$\boldsymbol{E} \cdot d\boldsymbol{S} = |\boldsymbol{E}|\,|d\boldsymbol{S}| = |\boldsymbol{E}|dS,$$

where in the dot product we have used the fact that the angle between the vector area of the segment and the electric field is zero. Furthermore, if the field depends only on $r$, then it will be constant on the surface of the

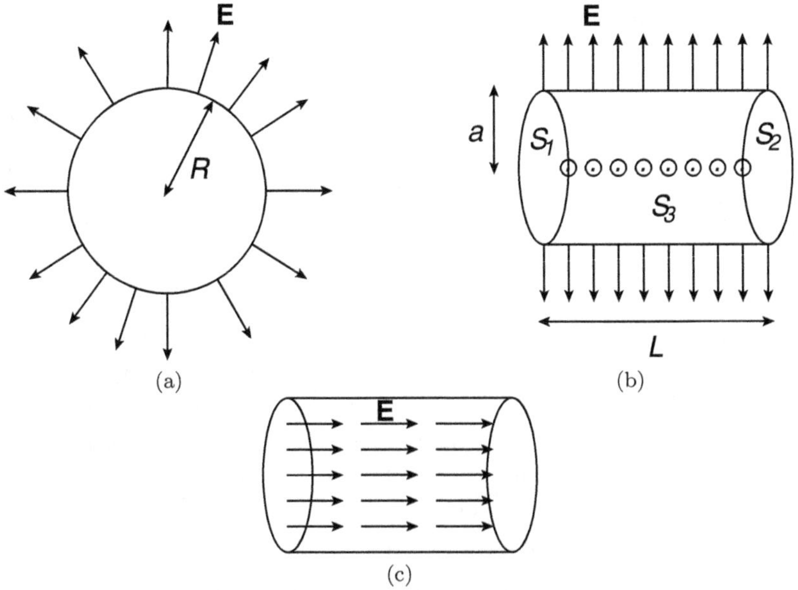


Fig. 3.12 (a) A spherical surface of radius $R$, with a radial electric field; (b) a cylindrical surface of radius $a$ and length $L$, with a field pointing radially outwards from the cylinder axis; (c) a cylindrical surface, with a uniform electric field parallel to the axis.

sphere, and the surface integral for the flux becomes

$$\Phi_E = \int\int_S \boldsymbol{E} \cdot d\boldsymbol{S} = \int\int_S |\boldsymbol{E}| dS = |\boldsymbol{E}| \int\int_S dS,$$

i.e. one can take the constant $|\boldsymbol{E}|$ outside the integral over the (scalar) area. Following our above remarks, the integral we are left with is to be interpreted as summing over the (scalar) areas $dS$ of each individual segment of the surface $S$. By definition, this is the total area of the surface, so that we have

$$\int\int_S dS = 4\pi R^2,$$

where we have used the standard result for the total area of a sphere. We thus find

$$\Phi_E = 4\pi R^2 |\boldsymbol{E}|,$$

without having had to do any integrals at all! Provided, of course, that we can remember the expression for the area of a sphere.

The above result makes sense: the flux is meant to describe the component of $\boldsymbol{E}$ which is perpendicular to a surface, times the area of a surface. If $|\boldsymbol{E}|$ is constant and *always* perpendicular to the surface, then even for a curved surface we can just write

$$\Phi_E = |\boldsymbol{E}| A, \tag{3.17}$$

where $A$ is the total area. We will also see examples which have *cylindrical* rather than spherical symmetry. An example is provided by Figure 3.12(b), which shows an electric field that points radially outwards from the axis of a cylinder. Let us also assume that it depends only on the perpendicular distance from this axis. Then consider the flux of $\boldsymbol{E}$ through the closed cylindrical surface shown. To calculate this, we can split the cylinder into three separate surfaces: $S_1$ and $S_2$ denote the endcaps, and $S_3$ the curved surface. On $S_1$ and $S_2$, $\boldsymbol{E}$ is parallel to the surface, and thus the contribution to the flux is zero (i.e. there is no component of $\boldsymbol{E}$ perpendicular to the surface, so no "stuff" crosses it). On $S_3$, the field is constant and perpendicular to the surface. Thus, similarly to the spherically symmetric case of Figure 3.12(a), one has

$$\Phi_E = |\boldsymbol{E}| A = 2\pi a L |\boldsymbol{E}|,$$

where $A$ is the area of the curved surface (we can get this by multiplying the circumference of the cylinder by its length).

As a further example, consider the same cylindrical closed surface, but with a uniform electric field pointing along the axis, shown in Figure 3.12(c). Now the contribution to the flux will be zero on the curved surface $S_3$, but non-zero on the endcaps $S_1$ and $S_2$. On $S_1$ we will have

$$\boldsymbol{E} \cdot d\boldsymbol{S} = |\boldsymbol{E}||d\boldsymbol{S}| \cos\theta,$$

where $\theta = 180°$ is the angle between the electric field and the vector area, where the latter points *outwards* from the surface. The cosine gives a minus sign:

$$\boldsymbol{E} \cdot d\boldsymbol{S} = -|\boldsymbol{E}|dS.$$

Given that the field is uniform, the flux through $S_1$ is

$$\Phi_E(S_1) = -|\boldsymbol{E}|A_1 = -\pi a^2|\boldsymbol{E}|,$$

where $A_1$ is the area of $S_1$. On $S_2$ the electric field is in the *same* direction as the vector area, and thus one has

$$\boldsymbol{E} \cdot d\boldsymbol{S} = |\boldsymbol{E}|dS,$$

which implies the flux through $S_2$ is

$$\Phi_E(S_2) = \pi a^2|\boldsymbol{E}|.$$

Combining the above results, the total flux through the entire surface is

$$\Phi_E(S_1) + \Phi_E(S_2) + \Phi_E(S_3) = 0.$$

Thus, this is a case where the flux is non-zero through parts of the surface, but the total flux is zero. The reason this can happen is that flux is *negative* if $\boldsymbol{E}$ flows into the surface, and *positive* if it flows out. In this particular case, equal amounts of flux flow in and out! Indeed, we know this just by looking at the picture, which justifies why such pictures are useful.

## 3.7 Gauss' Law

Armed with the concept of flux, we can derive an equation that allows us to find the electric field due to an arbitrary charge distribution. Consider first a positive charge $Q$ at the origin, and consider a spherical surface of

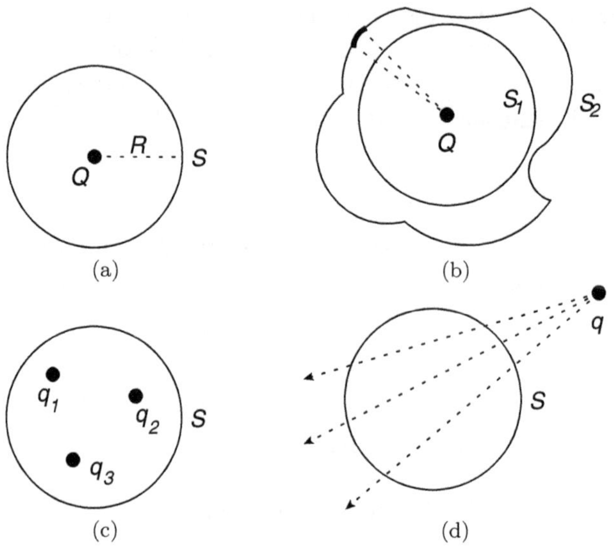


Fig. 3.13 (a) A charge $Q$ at the origin, surrounded by a spherical surface $S$; (b) a charge $Q$ inside an arbitrary surface $S_2$, which surrounds a spherical surface $S_1$; (c) several charges inside a spherical surface $S$; (d) a charge $q$ *outside* a spherical surface $S$.

radius $R$ surrounding it, as shown in Figure 3.13(a). From Coulomb's law, the magnitude of the electric field on the surface $S$ will be

$$|\boldsymbol{E}| = \frac{Q}{4\pi\epsilon_0 R^2},$$

which is constant, given that $R$ is constant on the sphere. From the results of the previous section, the total flux through the sphere will be given by

$$\Phi_E = |\boldsymbol{E}|A,$$

where $A = 4\pi R^2$ is the area of $S$. Explicitly, this is

$$\Phi_E = \frac{Q}{4\pi\epsilon_0 R^2}4\pi R^2 = \frac{Q}{\epsilon_0}.$$

Interestingly, $R$ has completely cancelled out of this expression. Thus, the flux is the same no matter what sphere we choose! In fact, we can make a statement considerably more general than this — the result turns out to be the same for *any* closed surface. To this end, consider the weirdly shaped surface $S_2$ in Figure 3.13(b), which also encloses a spherical surface $S_1$. If we zoom in finely enough on any part of the weirdly shaped surface, it will look locally like the surface of a sphere.[4] Given that we have seen that the radius of a sphere cancels out in the contribution to the flux, we can think of the surface $S_2$ as being composed of lots of little bits of sphere, so that the total flux is the same as if it *were* a sphere. Another way to think about this is to use the fluid analogy from when we first defined electric flux. The flux represents a "flow of stuff" across a given surface. It follows that a closed surface *captures all the stuff*, and that all of the stuff that flows through $S_1$ in Figure 3.13(b) must also eventually flow through $S_2$. Thus, the flux through $S_2$ is the same as the flux through a sphere.

Both of the situations just described had only a single charge living inside the closed surfaces being considered. We can now generalise our results further by asking what happens if more than one charge is present. As an example, Figure 3.13(c) shows three charges inside a surface that, as we have just seen, we may as well take to be spherical. The total flux through the surface is

$$\int\int d\boldsymbol{S}\cdot\boldsymbol{E} = \int\int d\boldsymbol{S}\cdot\sum_i \boldsymbol{E}_i,$$

where $\boldsymbol{E}_i$ is the electric field due to the charge $q_i$, and we have used the principle of superposition. We can now take the sum outside the integral (i.e. an integral of a sum of terms is the same as the sum of the integrals of the terms), to get

$$\begin{aligned}\int\int d\boldsymbol{S}\cdot\boldsymbol{E} &= \sum_i \int\int d\boldsymbol{S}\cdot\boldsymbol{E}_i\\ &= \sum_i \frac{q_i}{\epsilon_0}\\ &= \frac{1}{\epsilon_0}\sum_i q_i.\end{aligned}$$

In the second line, we have used the above result for the flux due to a single charge. It does not matter that each charge is not at the centre of the sphere, as we have seen that the flux is the same for any closed surface, which certainly includes a displaced sphere. In the final line, we

[4]You may be worried that some parts of a given surface will look like bits of a sphere that are curving outwards rather than inwards. However, upon zooming in further, all surfaces (including spheres) will look flat, so that the curvature can be ignored.

can recognise $\sum_i q_i$ as the total charge enclosed by the spherical surface. Writing this as $Q$, we can then write

$$\Phi_E = \int\int d\boldsymbol{S} \cdot \boldsymbol{E} = \frac{Q}{\epsilon_0},$$

i.e. the same result as before, but now we must take $Q$ to be the *total charge* enclosed by $S$.

Finally, let us ask if charges *outside* the surface $S$ also contribute to the flux. The answer to this is no, which can again be reasoned by using the fluid analogy. The flux represents a flow of stuff across a surface, and for the charge shown in Figure 3.13(d), this will be negative at the top of the sphere, as stuff is flowing *into* the surface. On the other hand, the flux is positive on the lower region of the sphere, as stuff is flowing *out* of the surface. These cancel out exactly, so that no net amount of stuff flows into or out of the sphere. A more detailed mathematical analysis can indeed be used to prove this properly.

To summarise the preceding discussion, we have seen that for *any* charge distribution, and for *any closed surface* $S$, one has

$$\oiint_S d\boldsymbol{S} \cdot \boldsymbol{E} = \frac{Q}{\epsilon_0}. \tag{3.18}$$

Note that I have used a peculiar symbol on the left, by putting a closed loop around the double integral symbol. This is the symbol for a *closed* surface integral, and can be useful as it reminds us that the above result only applies for closed surfaces. In words, Eq. (3.18) tells us that the electric flux through any closed surface is *always* equal to the total charge enclosed, divided by the permittivity. This result is usually called *Gauss' law*, and it is entirely equivalent in principle to Coulomb's law (i.e. we have used Coulomb's law explicitly in our derivation of the above result). However, note that Coulomb's law can be very cumbersome to apply. If we have a continuous charge distribution, or a set of multiple point charges in some weird configuration, it is usually much more difficult to use Coulomb's law than it is to simply use Gauss' law. The power of the latter in particular is that we can choose *any* closed surface in trying to find the flux which, as we will see, gives us considerable freedom in being able to quickly derive the result for the electric field in complicated situations.

We seem to have two different laws — Gauss' law and Coulomb's law — which ultimately amount to the same thing. So you may be wondering which is the more "fundamental" equation. This is partly a matter of choice of course, but the usual point of view is that we take Gauss' law of Eq. (3.18) to be a fundamental principle of nature, from which Coulomb's law is *derived.* Indeed, Gauss' law forms part of the Maxwell equations, as we will see. What also supports the view that this is the right thing to do is that Gauss' law is set up directly in terms of *the electric field* as the most important thing, rather than forces between individual charges.

The above discussion has all been very abstract, so let's look at a few examples of how to apply Gauss' law:

(1) *Rederiving Coulomb's law*: Imagine we did not know Coulomb's law. We could then derive it straight from Gauss' law. To this end, we want to find the field a distance $R$ from a given charge — once we have the field, we know the force a test charge would experience in the field. To find the field at distance $R$, we can use Gauss' law with any surface we like. Let us then put the charge $Q$ at the origin, and (given the spherical symmetry of the situation), use a spherical surface $S$ of radius $R$, centred on the point charge, as shown in Figure 3.13(a). By symmetry, the magnitude of the electric field is constant on the surface. Furthermore, because $\boldsymbol{E}$ points radially outwards from point charges, it points in the radial direction at all points on the sphere. The total flux will then be given by

$$\Phi_E = 4\pi R^2 |\boldsymbol{E}| = \frac{Q}{\epsilon_0},$$

where we have used Eq. (3.17). Rearranging then gives

$$|\boldsymbol{E}| = \frac{Q}{4\pi\epsilon_0 R^2},$$

which together with the direction precisely reproduces the electric field we obtained from Coulomb's law.

(2) *Field due to an infinite charged line*: Consider a line with uniform positive charge per unit length $\lambda$. We can then ask: what is the field a perpendicular distance $R$ from the wire? This can be found by applying Gauss' law to a cylindrical surface of radius $R$, where the cylinder is coaxial with the line. By symmetry, the field will point perpendicularly outwards from the cylinder, and this is precisely the situation considered in Figure 3.12(b). We may then use our previous result that the total flux will be the area of the curved part of the cylinder, times the (constant) magnitude of the electric field (recall that there is no flux

through the endcaps, as the field is parallel to the surface there). One thus finds

$$\Phi_E = |\boldsymbol{E}|2\pi RL = \frac{\lambda L}{\epsilon_0}.$$

The right-hand side follows from Gauss' law, and the fact that $\lambda L$ is the total charge enclosed by the cylinder, given that $\lambda$ is the charge per unit length along the line. The magnitude of the field is thus

$$|\boldsymbol{E}| = \frac{\lambda}{2\pi\epsilon_0 R}.$$

A useful check of this result is that it does not depend on $L$, which is the length of the *arbitrary* surface we chose when applying Gauss' law. It instead only depends on physically meaningful things such as the intrinsic properties of the wire ($\lambda$), how far we are from the wire ($R$), and the properties of the vacuum ($\epsilon_0$).

(3) *Infinite sheet of charge*: Consider an infinite sheet of charge, with constant positive charge per unit area $\sigma$. A section of this sheet is shown in Figure 3.14, and by symmetry the electric field on either side of the sheet will point in the vertical direction, and *away* from the sheet. Furthermore, it will only depend on the perpendicular distance from the plane. Let us then consider a cylindrical surface of height $h$ and radius $R$, such that this is bisected by the plane. The total flux through this surface is given by the flux through the curved side, and the flux

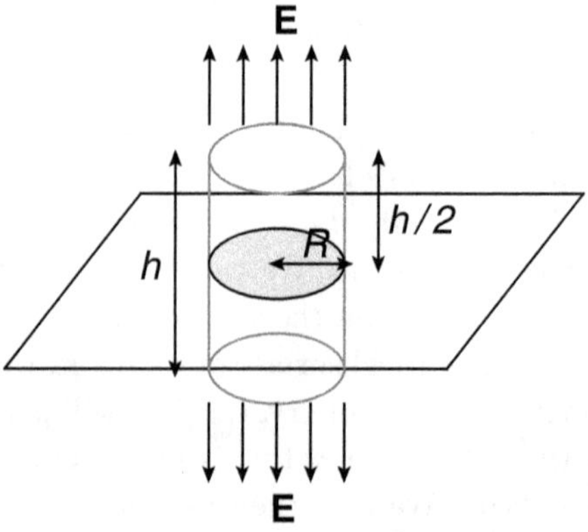


Fig. 3.14 An infinite charged sheet, with uniform charge $\sigma$ per unit area.

through the endcaps. The former is zero, as the electric field is everywhere parallel to the side of the cylinder. The flux through the upper endcap is

$$\pi R^2|\boldsymbol{E}|,$$

which is also the flux through the lower endcap, as in both cases the electric field points *outwards* from the surface. Hence, the total flux is

$$\Phi_E = 2\pi R^2|\boldsymbol{E}| = \frac{Q}{\epsilon_0} = \frac{\pi R^2\sigma}{\epsilon_0},$$

where on the right we have used that the total charge enclosed by the surface is the charge contained in the grey shaded area in Figure 3.14, namely

$$Q = \pi R^2\sigma.$$

Rearranging, the magnitude of the field is given by

$$|\boldsymbol{E}| = \frac{\sigma}{2\epsilon_0}.$$

Interestingly, we see that this result is entirely independent of $h$. That is, it does not decrease as we move further from the sheet. This is similar to a uniform gravitational field, a fact that we will revisit later.

(4) *The electric field near a conductor*: In a conductor, charged particles move freely, such that any external electric field is cancelled out. An example is shown in Figure 3.15(a), which shows a spherical conductor in an electric field $\boldsymbol{E}_a$. Mobile charges in the conductor will then move in the direction of the electric field (if they are positive), or against the field if they are negative. This leads to a build-up of charge in different

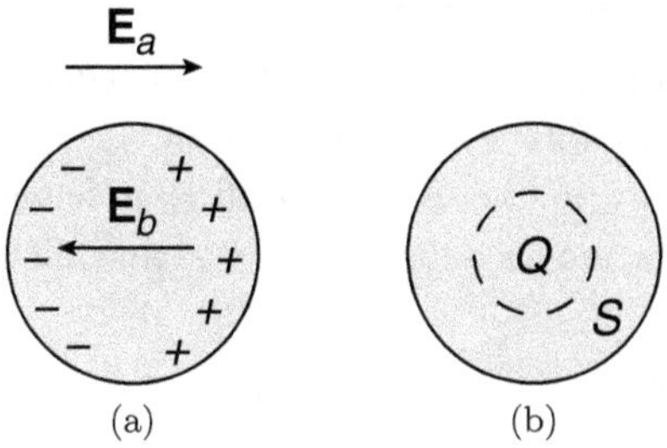


Fig. 3.15 (a) If an electric field $\boldsymbol{E}_a$ is applied to a conductor, charges move so as to create an *induced field* $\boldsymbol{E}_b$; (b) a spherical surface inside a conductor.

regions of the conductor, and the effect of this will be to create an *induced electric field* $\boldsymbol{E}_b$ inside the conductor. An equilibrium will be reached when this field precisely cancels out the applied field $\boldsymbol{E}_a$, as at this point there will be no force on the charges. That is, inside the conductor the total electric field is

$$\boldsymbol{E} = \boldsymbol{E}_a + \boldsymbol{E}_b = \mathbf{0}.$$

Similarly, $\boldsymbol{E}$ at the surface of the conductor must be purely perpendicular to the surface: any component parallel to the surface would lead to movement of charges, which would build up so as to cancel this component. We can also conclude that any build-up of excess charge *must* reside on the surface of a conductor. Consider applying Gauss' law to a surface inside a conductor, as in Figure 3.15(b). Gauss' law gives

$$\int\int d\boldsymbol{S} \cdot \boldsymbol{E} = 0 = \frac{Q}{\epsilon_0},$$

where in the first equality we have used the fact (derived above) that the electric field inside the conductor is zero. It follows that the net charge inside the conductor is zero, such that any excess charge must reside on the surface only.

This example shows the full power of Gauss' law. Conducting materials are very complicated systems in general, full of an enormous number of positive and negative charges. It is clearly not possible to analyse such a system using Coulomb's law alone. Instead, Gauss' law coupled with some simple assumptions about how conductors behave (which are really just the definition of what a conductor is) allows us to derive some very general properties.

## 3.8 Electric Potential Energy

Whenever we have a force, we can talk about a *potential energy* associated with the force. Electric forces are no exception, so let's see how we can talk about potential energy in this case. Recall that when a uniform force $\boldsymbol{F}$ acts on a particle that moves a distance $l$, we can talk about the *work done* by the force:

$$W = |\boldsymbol{F}|l. \tag{3.19}$$

We can also generalise this definition to cases in which the distance moved is not in the direction of the force, as shown in Figure 3.16(a). Now the

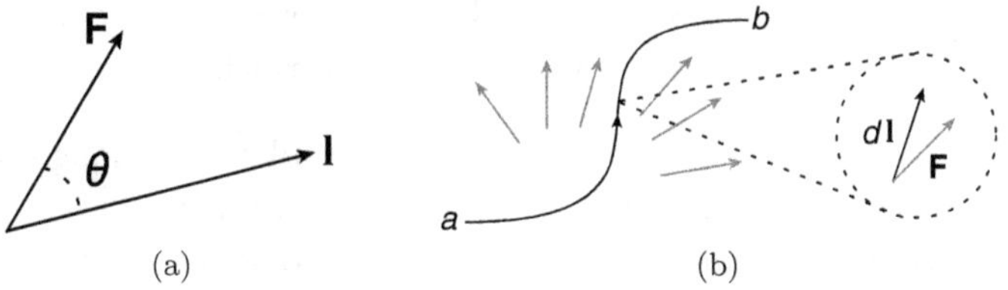


Fig. 3.16 (a) A uniform force $\boldsymbol{F}$ at angle $\theta$ to a displacement $\boldsymbol{l}$; (b) a curved path from $a$ to $b$, in a non-uniform force field $\boldsymbol{F}(\boldsymbol{x})$.

distance moved *in the direction of the force* is $|\boldsymbol{l}|\cos\theta$, so that the work done is

$$W = |\boldsymbol{F}||\boldsymbol{l}|\cos\theta = \boldsymbol{F}\cdot\boldsymbol{l},$$

where we have recognised the dot product in the second equality.

This is still not the most general situation. In principle, one can have a nonuniform field of force $\boldsymbol{F}(\boldsymbol{x})$, such that the force is different at every point in space. Also, we might move a particle on an arbitrary curved path between two points $a$ and $b$, rather than on a straight line. This is shown in Figure 3.16(b), where the curve is shown in black, and the force field in grey. What we can then do is zoom in to a small segment of the curve, such that it looks like a straight line, represented by the displacement vector $d\boldsymbol{l}$. That is, $d\boldsymbol{l}$ points along the curve, and has a magnitude $dl \equiv |d\boldsymbol{l}|$, equal to the small distance traversed along the curve. Then

$$L = \int dl$$

is the total length of the curve. Provided $d\boldsymbol{l}$ is infinitesimally small, the force $\boldsymbol{F}$ will be constant along this displacement, so that the contribution to the work done by the force is

$$dW = \boldsymbol{F}\cdot d\boldsymbol{l},$$

where the force is evaluated at the segment $d\boldsymbol{l}$. The total work done is then

$$W = \int_a^b \boldsymbol{F}\cdot d\boldsymbol{l}, \tag{3.20}$$

where the limits remind us that the curve goes between points $a$ and $b$. Equation (3.20) is called a *line integral*, and can be quite complicated for general paths, or for complicated space-dependent force fields $\boldsymbol{F}(\boldsymbol{x})$.

We discuss the full complication of how to carry out line integrals in appendix A. However, there are two extreme cases where things simplify:

(i) The force $\boldsymbol{F}$ is *perpendicular* to the curve at all points along the curve. Then $\boldsymbol{F} \cdot d\boldsymbol{l} = 0$ on all segments $d\boldsymbol{l}$, and no work is done.
(ii) The force $\boldsymbol{F}$ is *parallel* to $d\boldsymbol{l}$ at all points along the curve. Then one has

$$\boldsymbol{F} \cdot d\boldsymbol{l} = |\boldsymbol{F}| dl,$$

and the line integral becomes a normal scalar integral.

We will see many examples of such integrals later, but for now consider the simple situation of a uniform field in the $x$ direction, as shown in Figure 3.17(a). Imagine that a particle is moved from $a$ to $b$ along the straight-line contour shown. On this curve, we can write

$$dl = dx, \quad a \leq x \leq b.$$

That is, $dx$ indeed represents the infinitesimal distance at an arbitrary point along the curve, such that the integral from $a$ to $b$ is the total length. One then has

$$\int \boldsymbol{F} \cdot d\boldsymbol{l} = \int_a^b |\boldsymbol{F}| dx = |\boldsymbol{F}| \int_a^b dx = |\boldsymbol{F}| l,$$

where we have recognised $l = b - a$ as the length of the curve, and also used that $|\boldsymbol{F}|$ is constant. The final result agrees with the previous definition of the work done by a uniform force, if a particle is moved a distance $l$ along the direction of the force.

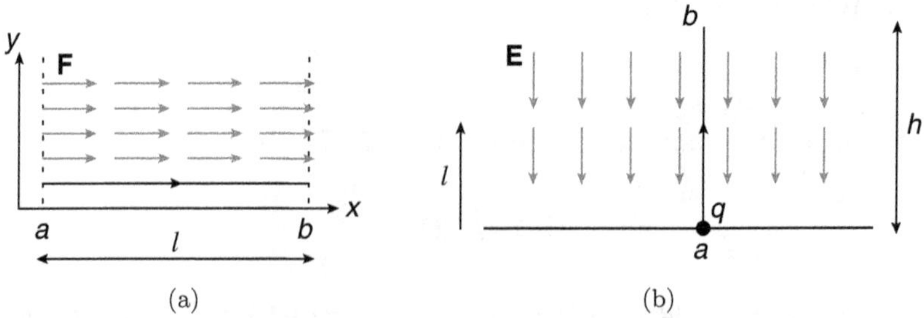


Fig. 3.17 (a) A uniform force field in the $x$ direction; (b) a uniform electric field in the vertical direction, where the charge $q$ is moved along the path shown.

Now recall that by the conservation of energy, any work done by a force must be balanced by the change in the *potential energy* of the moving particle, where different potential energies are associated with different forces. That is, we may write

$$W = -\Delta U, \tag{3.21}$$

where $\Delta U$ is the change in potential energy experienced by the particle. Note that care is needed here with signs: if the particle moves *against* the force, its potential energy *increases* (e.g. think of the case of a particle being lifted in a uniform gravitational field). The work done by the force in this case is negative (by its definition), hence the minus sign above. Combining Eqs. (3.20) and (3.21), we get

$$\Delta U = -\int_a^b \boldsymbol{F} \cdot d\boldsymbol{l}. \tag{3.22}$$

Let us now consider the specific case of the electric force. A given particle of charge $q$ experiences a force

$$\boldsymbol{F} = q\boldsymbol{E},$$

and thus has *electric potential energy* $U_E$, such that the change in this potential energy in moving the particle from $a$ to $b$ along a given path is given by

$$\Delta U_E = -q\int_a^b \boldsymbol{E} \cdot d\boldsymbol{l}. \tag{3.23}$$

For example, consider the uniform electric field of Figure 3.17(b), pointing in the downwards direction, and imagine moving a charge $q$ from the ground to a height $h$, along the vertical path shown. We can then define the distance $l$ as increasing upwards from the ground, so that one has

$$\begin{aligned} \Delta U_E &= -q\int_a^b \boldsymbol{E} \cdot d\boldsymbol{l} \\ &= q\int_a^b |\boldsymbol{E}| dl, \end{aligned}$$

where in the second line we have used the fact that $d\boldsymbol{l}$ on each segment of the curve is pointing in the opposite direction to the field $\boldsymbol{E}$, so that

$$\boldsymbol{E} \cdot d\boldsymbol{l} = -|\boldsymbol{E}||d\boldsymbol{l}| = -|\boldsymbol{E}|dl.$$

Using the fact that $|\boldsymbol{E}|$ is constant, we may rewrite the change in potential energy as

$$\Delta U_E = q|\boldsymbol{E}| \int_0^h dl \tag{3.24}$$

$$= q|\boldsymbol{E}|h. \tag{3.25}$$

Interestingly, this looks very like the change in gravitational potential energy upon lifting an object of mass $m$ a height $h$ in a uniform gravitational field $\boldsymbol{g}$:

$$\Delta U_{\text{grav.}} = m|\boldsymbol{g}|h. \tag{3.26}$$

Comparing Eqs. (3.25) and (3.26), we see that charge $q$ in electromagnetism plays the same role as mass $m$ in gravity, if $\boldsymbol{E}$ plays a similar role to $\boldsymbol{g}$. Relations between electromagnetism and gravity go much deeper than this, as we will explore in Chapter 10.

## 3.9 Electric Potential

The electric field represents the force that would be experienced by a test charge $q$, divided by the charge. It is thus independent of the details of a specific test charge, and it is useful to define a similar quantity relating to potential energy. Namely one defines the *electric potential*

$$V = \frac{U_E}{q}, \tag{3.27}$$

where $U_E$ is the potential energy that a given test charge $q$ would experience in a given electric field. The SI units of this quantity are $\text{JC}^{-1}$ (Joules per Coulomb). Like potential energy, the electric potential is not unique: one can choose where to define $V = 0$ (e.g. think again of gravitational potential energy, where one can choose it to be zero at the ground for convenience). The reason such an ambiguity is possible is that all physical phenomena depend only on *differences* in potential energy. Likewise, all observable electric phenomena depend only on *potential differences* $\Delta V$. Given Eqs. (3.23) and (3.27), we find that we can define the potential difference, from $a$ to $b$ along a given contour, via the line integral

$$\Delta V = -\int_a^b \boldsymbol{E} \cdot d\boldsymbol{l}. \tag{3.28}$$

Given that $\boldsymbol{E}$ is independent of the test charge $q$, so is $\Delta V$. Potential differences are usually quoted in *volts* V. The units of electric field can then be given as $\mathrm{Vm}^{-1}$ (volts per metre).

Note from the above definitions that the work done by a force in moving a charge $Q$ through a potential difference $\Delta V$ is

$$W = -Q\Delta V.$$

Equation (3.28) suggests that we can find the potential in any given situation once we know the electric field, and we now consider some examples:

(1) *Potential due to a point charge*: We have already seen that, for a point charge,

$$\boldsymbol{E} = \frac{Q}{4\pi\epsilon_0 r^2}\hat{\boldsymbol{r}},$$

where $\hat{\boldsymbol{r}}$ is a unit vector in the radial direction. By spherical symmetry, the potential can only depend on $r$. Consider moving a unit postive test charge from $r = r_B$ to $r = r_A$, along the radial path shown in Figure 3.18. On this path, we have

$$d\boldsymbol{l} = dr\hat{\boldsymbol{r}},$$

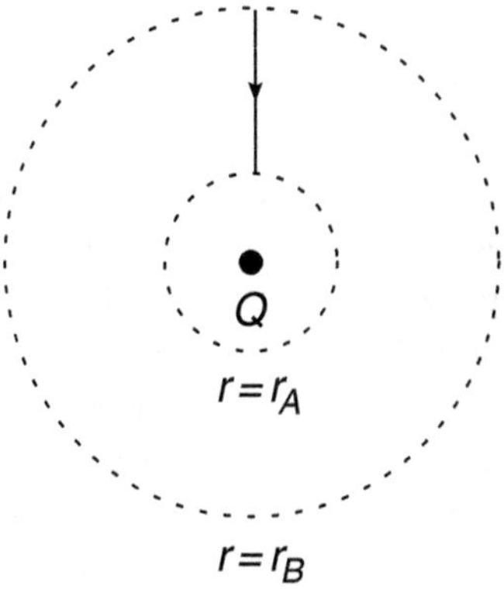


Fig. 3.18 A radial path from $r = r_B$ to $r = r_A$, where $r_B > r_A$.

where the unit radial vector $\hat{\boldsymbol{r}}$ points outwards from the origin. However, $r$ is decreasing on the path, so that $dr < 0$. Thus, the vector $\boldsymbol{dl}$ points *towards* the origin as required. Then

$$\boldsymbol{E} \cdot \boldsymbol{dl} = \frac{Q dr}{4\pi\epsilon_0 r^2} \hat{\boldsymbol{r}} \cdot \hat{\boldsymbol{r}} = \frac{Q dr}{4\pi\epsilon_0 r^2},$$

which in turn implies

$$\begin{aligned}\Delta V &= -\int \boldsymbol{E} \cdot \boldsymbol{dl} \\ &= -\int_{r_B}^{r_A} dr \frac{Q}{4\pi\epsilon_0 r^2} \\ &= \frac{Q}{4\pi\epsilon_0}\left(\frac{1}{r_A} - \frac{1}{r_B}\right).\end{aligned}$$

Often you will see people talk simply about the *electric potential*, without explicitly stating that this is a potential difference. Implicit in such definitions is always the fact that one has made a choice about where the potential is zero, so that all other values of the potential are evaluated relative to that. A common choice for the electric potential is to choose this to be zero at infinity. Then we can define $V(r)$ to be the potential difference on bringing a test charge from infinity to a distance $r$ from the point charge $Q$. In the above calculation this amounts to setting $r_A = r$, and taking $r_B \to \infty$, to give

$$V(r) = \frac{Q}{4\pi\epsilon_0 r}. \tag{3.29}$$

Note that this is positive for $Q > 0$, and a rough sketch is shown in Figure 3.19(a). The effect this has on a test particle can be thought about using the fact that objects want to minimise their potential energy, and thus the potential in this case. If we place a ball on the potential curve (representing a positive test charge), the direction it rolls in then gives us the direction of decreasing potential. This is to the right in Figure 3.19(a), which has the effect of increasing $r$. This is indeed correct: positive charges *repel* other positive charges, such that they move towards larger $r$ values. If we had a negative charge $Q < 0$ at the origin, the potential curve would instead look like Figure 3.19(b). A positive test charge would then roll to the left, towards lower $r$ values: negative charges *attract* positive test charges, as expected.

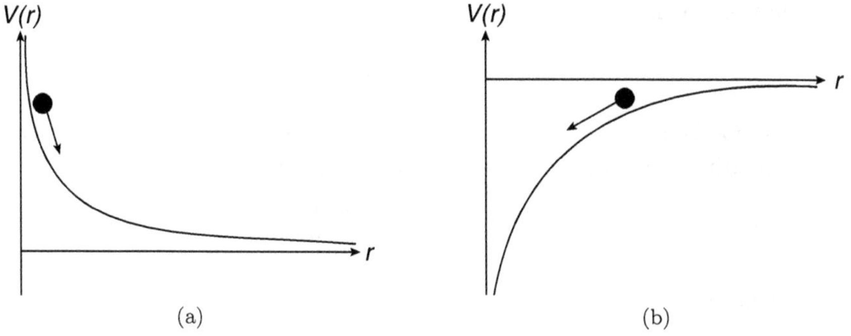


Fig. 3.19 The potential due to (a) a positive point charge at the origin; (b) a negative point charge. In each case the way a ball would roll down the curve gives us the direction of the force.

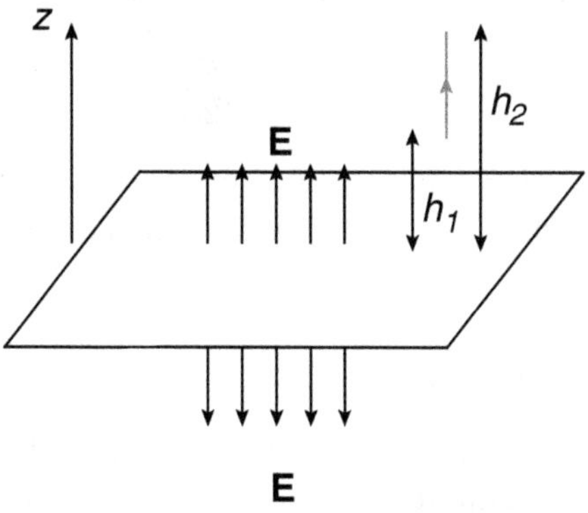


Fig. 3.20 Path (shown in grey) of a test charge near an infinitely charged sheet.

(2) *Infinite sheet of charge*: Earlier we found that the magnitude of the electric field a perpendicular distance $h$ from an infinite sheet of charge is given by

$$|\boldsymbol{E}| = \frac{\sigma}{2\epsilon_0},$$

where $\sigma$ is the charge per unit area of the sheet, and the direction of the field is outwards from the sheet if $\sigma > 0$. Now consider moving a particle from a height $h_1$ above the sheet to a height $h_2$, along a straight line path upwards from the sheet, as shown in Figure 3.20.

On this path, one has

$$d\boldsymbol{l} = dz\hat{\boldsymbol{k}},$$

where $\hat{\boldsymbol{k}}$ is a unit vector in the $z$ direction. The electric field is in the *same* direction as the path (and hence $d\boldsymbol{l}$), thus one has

$$\boldsymbol{E} \cdot d\boldsymbol{l} = dz\frac{\sigma}{2\epsilon_0},$$

which in turn implies

$$\Delta V = -\int_{h_1}^{h_2} dz\frac{\sigma}{2\epsilon_0} = \frac{\sigma}{2\epsilon_0}(h_1 - h_2).$$

We can choose to define $V = 0$ at $h = 0$, in which case the electric potential as a function of height $h$ above the sheet becomes

$$V(h) = -\frac{\sigma h}{2\epsilon_0}.$$

This is negative for $\sigma > 0$, so that positive test particles move to larger values of $h$, as expected given that they should follow the direction of the electric field.

## 3.10 Conservative Fields

If you think carefully, you may realise that I was not completely careful when defining the potential difference in Eq. (3.28). Looking at the limits of integration on the right-hand side, I chose to specify only the end points of the curve that the particle moves along. However, this is not enough to specify the curve, as there are infinitely many curves with the same endpoints! If the integral on the right depends on the precise details of the curve other than its endpoints, then we are in trouble. In fact, there is no reason to worry, which can be justified as follows. We have seen that the electric force depends only on position, and this in turn means that the potential energy depends only on position, so that if we take a test particle around a closed loop, the total work done must be zero. Phrased in terms of the electric field (i.e. the force divided by the charge on a test particle), this implies

$$\int_{C_1} \boldsymbol{E} \cdot d\boldsymbol{l} + \int_{C_2} \boldsymbol{E} \cdot d\boldsymbol{l} = 0,$$

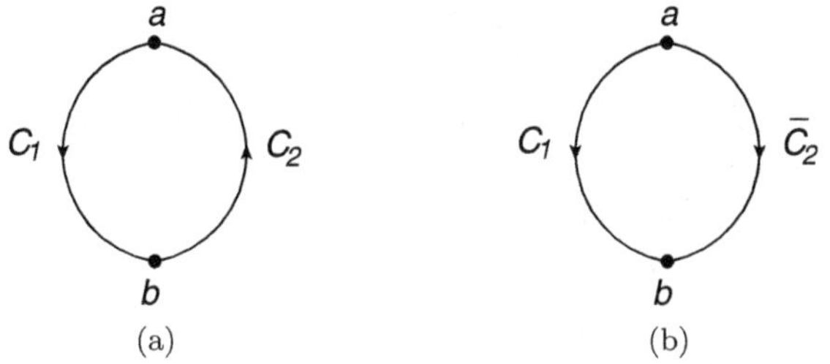


Fig. 3.21 (a) A closed loop can be decomposed into two contours $C_1$ and $C_2$ between points $a$ and $b$, joined together as shown; (b) reversing contour $C_2$ to get an alternative contour $\bar{C}_2$.

where $C_1$ and $C_2$ are the contours shown in Figure 3.21(a), and such that $C_1$ followed by $C_2$ constitutes a closed loop. We may instead interpret this equation differently, by reversing the contour $C_2$, and we label this by $\bar{C}_2$ in Figure 3.21(b). At each point on $\bar{C}_2$, the vector $\boldsymbol{dl}$ denoting the differential displacement is *reversed* with respect to its equivalent segment on $C_2$, so that we must have

$$\int_{\bar{C}_2} \boldsymbol{E} \cdot \boldsymbol{dl} = -\int_{C_2} \boldsymbol{E} \cdot \boldsymbol{dl}.$$

Thus, the above condition on the electric field may be rewritten as

$$\int_{C_1} \boldsymbol{E} \cdot \boldsymbol{dl} = \int_{\bar{C}_2} \boldsymbol{E} \cdot \boldsymbol{dl}.$$

Note that we have not at all specified the precise contours $\bar{C}_1$ and $\bar{C}_2$: they can be *any* contours joining the points $a$ and $b$. Hence we arrive at the following:

*The line integral of the electric field between a and b is independent of the path taken.*

Forces with this property are called *conservative forces*, and have the following (equivalent) properties:

(1) The line integral (work done) around a closed loop is zero.
(2) The line integral (work done) is independent of the path taken between two given points $a$ and $b$.
(3) There is a well-defined position-dependent potential.
(4) The work done is *reversible*, i.e. no energy is lost (hence the word "conservative").

Condition 2 follows directly from condition 1, as we have seen above. For condition 3, recall that we defined the potential $V(\boldsymbol{x})$ to be the energy difference (per unit charge) upon bringing a test charge from infinity to a point $\boldsymbol{x}$. This concept itself only makes sense if the path chosen is irrelevant. Furthermore, condition 4 follows from condition 1: if energy is lost upon taking a test charge around a closed loop, the work done would be non-zero.

## 3.11 Superposition of Potential

Earlier we saw that a superposition principle applied to the electric field $\boldsymbol{E}$. A similar principle applies also to the potential $V(\boldsymbol{x})$, which gives an alternative to calculating this from the electric field. As before, let $V(r)$ be the potential difference upon bringing a positive test charge from infinity to a point $a$ which is a distance $r$ from a point charge $Q$:

$$V(r) = \frac{Q}{4\pi\epsilon_0 r}, \quad V = -\int_\infty^a d\boldsymbol{l} \cdot \boldsymbol{E}.$$

Now consider a system of $N$ charges $\{q_i\}$. Superposition for the electric field gives

$$\boldsymbol{E} = \sum_{i=1}^{N} \boldsymbol{E}_i,$$

where $\boldsymbol{E}_i$ is the electric field due to the charge $i$. Then the potential at the point $a$ for this system of charges is given by

$$\begin{aligned} V &= -\int_\infty^a \boldsymbol{E} \cdot d\boldsymbol{l} \\ &= -\int_\infty^a \left(\sum_{i=1}^{N} \boldsymbol{E}_i\right) \cdot d\boldsymbol{l} \\ &= -\sum_{i=1}^{N} \int_\infty^a \boldsymbol{E}_i \cdot d\boldsymbol{l} \\ &= \sum_{i=1}^{N} V_i, \end{aligned}$$

where $V_i$ is the contribution to the potential from charge $q_i$, as if there were no other charges present. Note what has happened here — the fact that the potential is *linear* in the electric field means that superposition for the electric field translates to superposition for the potential.

A similar argument can be made for a continuous charge distribution. As an example, we may consider the uniformly charged ring of Figure 3.7, with total charge $Q$ and radius $a$, and ask: what is the potential at a point $x$ along the $x$-axis? To find this, we can note that each segment of the ring with charge $dq$ contributes to the potential according to

$$dV = \frac{dq}{4\pi\epsilon_0 r} = \frac{dq}{4\pi\epsilon_0(x^2+a^2)^{1/2}},$$

where we have applied Pythagoras' theorem in the second equality. The total potential is then given by superposition of the potential due to each individual segment, namely the integral

$$V = \int \frac{dq}{4\pi\epsilon_0(x^2+a^2)^{1/2}} = \frac{1}{4\pi\epsilon_0(x^2+a^2)^{1/2}} \int dq = \frac{Q}{4\pi\epsilon_0(x^2+a^2)^{1/2}}.$$

At large distances ($x \gg a$) this looks like the potential of a point charge, as it must do given that one cannot then resolve the structure of the ring.

We have here talked about the potential $V(\boldsymbol{x})$ as referring to the potential difference experienced by a test charge upon being brought from infinity to the point $\boldsymbol{x}$. That is, we say that the potential $V = 0$ on the spherical surface at $r = \infty$. One could instead have chosen a different surface on which to define $V = 0$. As the above derivation of superposition for the potential hopefully makes clear, one can only superpose potentials if they have been defined with respect to the *same* surface, so please be careful!

## 3.12 Potential Energy of a System of Charges

The above result for the potential due to a system of charges (e.g. the system in Figure 3.4) is the total potential experienced by a test charge, in the system of charges $\{q_i\}$. Accordingly, there is a potential energy upon bringing a given point charge in towards the system of charges. It follows that the system of charges itself also has a potential energy associated with it, where this potential energy arises from the interaction of the charges with each other. We can find this by starting with no charges, and then bringing all the charges together from infinity, one by one.

The first charge $q_1$ will feel no potential, as there are not yet any other charges present. Then, charge $q_2$ will feel the potential

$$\frac{q_1}{4\pi\epsilon_0 r_{12}}$$

due to charge $q_1$, where we let $r_{ij}$ denote the distance between the charges $q_i$ and $q_j$. Continuing with this procedure, charge $q_3$ will feel a potential

$$\frac{q_1}{4\pi\epsilon_0 r_{13}} + \frac{q_2}{4\pi\epsilon r_{23}},$$

due to its interactions with both charges $q_1$ and $q_2$. Continuing with this procedure, we can see that upon bringing in charge $q_i$ from infinity, it will experience a potential due to the interaction with all charges $q_j$, with $j < i$:

$$\sum_{j<i} \frac{q_j}{4\pi\epsilon_0 r_{ji}}.$$

Given that the potential is defined to be the potential energy per unit charge (Eq. (3.27)), the potential energy experienced by the charge $q_i$ is

$$q_i \sum_{j<i} \frac{q_j}{4\pi\epsilon_0 r_{ji}},$$

so that the total potential energy upon assembling all $N$ charges is

$$U = \sum_{i=1}^{N} \sum_{j<i} \frac{q_i q_j}{4\pi\epsilon_0 r_{ji}}.$$

It is possible to write this in a more symmetric form. First of all, given that $r_{ji}$ represents the distance between charges $q_i$ and $q_j$, we can write

$$r_{ij} = r_{ji}.$$

This in turn means that the argument inside the double sum above is completely symmetric upon interchanging $i$ and $j$. We can then write

$$2U = \sum_{i=1}^{N} \left[ \sum_{i<j} \frac{q_i q_j}{4\pi\epsilon_0 r_{ji}} + \sum_{i>j} \frac{q_i q_j}{4\pi\epsilon_0 r_{ji}} \right].$$

That is, the set of all pairs $(q_i, q_j)$ with $i < j$ is the same as the set with $i > j$, after relabelling $i$ and $j$. This does not affect the argument of the sum, so that including both $i < j$ and $i > j$ on the right-hand side simply means that we have multiplied the potential energy by a factor of two. The set of all ways that $i$ can be bigger then $j$, or $j$ can be bigger than $i$, is equivalent to the number of ways in which $i$ is *not equal* to $j$, so that we can rewrite

$$\sum_{i<j} \frac{q_i q_j}{4\pi\epsilon_0 r_{ji}} + \sum_{i>j} \frac{q_i q_j}{4\pi\epsilon_0 r_{ji}} = \sum_{j\neq i} \frac{q_i q_j}{4\pi\epsilon_0 r_{ji}}.$$

Finally, one obtains

$$U = \frac{1}{2} \sum_{i=1}^{N} \sum_{j\neq i} \frac{q_i q_j}{4\pi\epsilon_0 r_{ji}}.$$

An alternative way to consider this result is to note that

$$V_i = \sum_{j\neq i} \frac{q_j}{4\pi\epsilon_0 r_{ji}}$$

is the total potential of charge $i$ due to all the other charges. The above result then becomes

$$U = \frac{1}{2} \sum_{i=1}^{N} q_i V_i. \tag{3.30}$$

That is, to find the total potential energy of a system of charges, one multiplies each charge by its potential, and then divides by two. The above derivation is admittedly quite horrible, especially given that the end result is quite simple. In particular, the factor of two in the final result perhaps still feels a bit mysterious, so let us give a simpler way to understand it. First, note that the each individual pair of charges has a potential energy associated with it (due to each charge sitting in the electric field of the other one). Thus, the total potential energy should be due to the sum of potential energies due to all *distinct* pairs of charges. If we were to naïvely form the sum

$$\sum_{i=1} q_i V_i,$$

we would have counted each pair of charges twice. We thus need to divide by two to correct for this overcounting.

The above result generalises to continuous charge distributions. Let $\rho(\boldsymbol{x})$ be the charge per unit volume, such that the charge in a small volume $dV = dxdydz$ is

$$dq = \rho(\boldsymbol{x})dxdydz,$$

with potential $V(\boldsymbol{x})$. The potential energy of such a continuous distribution is then obtained by summing over each small volume, to get

$$U = \frac{1}{2}\sum_{i=1} q_i V_i \to \frac{1}{2}\int dx \int dy \int dz \rho(\boldsymbol{x}) V(\boldsymbol{x}), \tag{3.31}$$

or simply

$$U = \frac{1}{2}\int dq V(\boldsymbol{x}). \tag{3.32}$$

Given this potential energy, it is natural to ask where it resides. Our modern understanding of this is that it is stored within the electric field itself! That is, for a given electric field $\boldsymbol{E}(\boldsymbol{x})$, we can associate a given *energy density* (potential energy per unit volume). We can arrive at the correct result by considering a specific example. Consider the two infinite uniformly charged planes shown in Figure 3.22 and let the upper and lower planes carry a charge per unit area $\pm\sigma$, respectively, where $\sigma > 0$. We saw earlier that for a single plane, the electric field points outwards from the plane (for positive charge), and towards the plane (for negative charge). The magnitude of the electric field is uniform and given by

$$|\boldsymbol{E}_{\text{single}}| = \frac{\sigma}{2\epsilon_0}.$$

If we consider the region between the two planes in Figure 3.22, the electric fields due to the upper and lower planes will have the same magnitude, and will be pointing in the *same* direction. Thus, they can be superposed

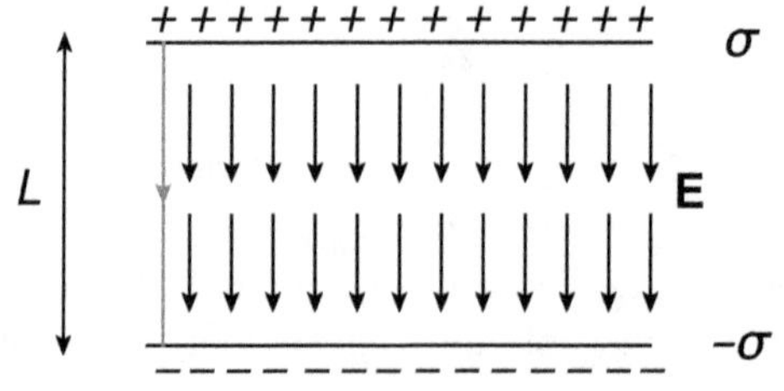


Fig. 3.22 A section of two infinitely charges planes separated by a distance $L$, where the upper and lower planes carry charge per unit area $\pm\sigma$, respectively.

to give

$$|\boldsymbol{E}| = \frac{\sigma}{\epsilon_0}.$$

Given that we can choose where to set the zero of potential, let us take this to be zero on the upper plane. Then the potential on the lower plane is given by

$$V = -\int_C \boldsymbol{E} \cdot d\boldsymbol{l},$$

where $C$ is any contour that goes from the upper plane to the lower plane. The simplest thing to choose is a vertical line as shown in the figure. Given that the field is uniform we simply get

$$V = -\frac{\sigma L}{\epsilon_0}.$$

We can now find the potential energy stored by the electric field in between the planes, by using Eq. (3.32). Given that we have taken $V = 0$ on the upper plane, the only place that the product of the potential and the charge is non-zero is on the lower plane. Consider a small area $dS$ on this plane. On this segment, one has

$$V dq = V(-\sigma) dS = \frac{\sigma^2 L}{\epsilon_0} dS.$$

The total potential energy for a given communal area $A$ of the planes is then

$$U = \frac{1}{2}\int V dq = \frac{1}{2}\int\int dS \frac{\sigma^2 L}{\epsilon_0} = \frac{\sigma^2 LA}{2\epsilon_0}.$$

The product $LA$ constitutes a fixed volume in between the planes, and we can therefore write the energy per unit volume between the planes as

$$\frac{U}{LA} = \frac{\sigma^2}{2\epsilon_0} = \frac{\epsilon_0}{2}\left(\frac{\sigma}{\epsilon_0}\right)^2 = \frac{1}{2}\epsilon_0 |\boldsymbol{E}|^2. \tag{3.33}$$

This is a very special case, but the result turns out to be very general, for arbitrary electric fields:

*The energy density of the electric field is given by* $\frac{1}{2}\epsilon_0|\boldsymbol{E}|^2$.

Note in particular that, because we are talking about an energy *density* which is a property of the field at a single point, it does not matter if the electric field is varying with position. This result is of practical importance — there are many cases where we might want to know how much energy is being stored by an electric field (or how much energy is needed to produce one). However, I hope you can appreciate that this result is also conceptually interesting. Earlier on, I said that our modern understanding of electromagnetism has the fields playing a central role. For this to make sense, they must be physically real in some sense, or at least as real as anything else we talk about in physics (e.g. particles like the electron and proton). One of the characteristics of "real" things is that they can carry energy, and so it is reassuring to see that our electric field — which looked like a mere mathematical convenience when first introduced — is starting to come alive! We can also go further than this, if you already know about Special Relativity. Energy and momentum mix up in that theory. Thus, by saying that the electric field can carry energy, it follows that it can also carry momentum in general.

## 3.13 Electric Field as the Gradient of Potential

We have seen that the potential is related to the electric field by the line integral:

$$V = -\int_{\mathcal{C}} \boldsymbol{E} \cdot d\boldsymbol{l},$$

where $\mathcal{C}$ is the chosen contour. It is useful to know what the inverse of this relationship is: given a particular potential function, can we find the corresponding electric field? To see how to do this, note that a given segment of the path $\mathcal{C}$ has

$$d\boldsymbol{l} = \begin{pmatrix} dx \\ dy \\ dz \end{pmatrix},$$

for some $dx$, $dy$ and $dz$. Then

$$\boldsymbol{E} \cdot dl = E_x dx + E_y dy + E_z dz,$$

which in turn implies that the change in potential upon traversing the segment $d\boldsymbol{l}$ is

$$dV = -E_x dx - E_y dy - E_z dz.$$

It follows immediately from this expression that

$$E_x = -\frac{\partial V}{\partial x}, \quad E_y = -\frac{\partial V}{\partial y}, \quad E_z = -\frac{\partial V}{\partial z},$$

where in each case the partial derivative implies that the other coordinates should be held constant. More succinctly, one has

$$\boldsymbol{E} = \begin{pmatrix} -\partial V/\partial x \\ -\partial V/\partial y \\ -\partial V/\partial z \end{pmatrix} = -\begin{pmatrix} \partial V/\partial x \\ \partial V/\partial y \\ \partial V/\partial z \end{pmatrix}.$$

There is a truly ingenious notation that helps us to remember this, and that is used in almost every textbook you will come across. Let us define the symbol (pronounced "nabla")

$$\nabla \equiv \begin{pmatrix} \partial/\partial x \\ \partial/\partial y \\ \partial/\partial z \end{pmatrix}. \tag{3.34}$$

This is a strange quantity: it is not quite a vector itself, and instead only makes sense if it operates on a function. It is thus called a *vector operator*, and acts on scalar functions to produce vectors. With this notation, the electric field can be rewritten as

$$\boldsymbol{E} = -\nabla V. \tag{3.35}$$

Provided we can remember the definition of Eq. (3.34), the notation does the work for us of remembering the components of the electric field. The operation appearing in Eq. (3.35) is referred to as the *gradient* (i.e. $\boldsymbol{E}$ is minus the gradient of $V(\boldsymbol{x})$), and in old books may be written as

$$\boldsymbol{E} = -\text{grad}V.$$

As an example, we may again consider the charged ring of Figure 3.7, for which we have previously found the potential and electric field at an arbitrary point a distance $x$ along the $x$-axis to be

$$V = \frac{Q}{4\pi\epsilon_0(x^2 + a^2)^{1/2}}, \quad \boldsymbol{E} = \frac{Q}{4\pi\epsilon_0} \frac{x}{(x^2 + a^2)^{3/2}} \hat{\boldsymbol{\imath}},$$

where $Q$ and $a$ are the total charge and radius of the ring, respectively, and $\hat{\boldsymbol{\imath}}$ the unit basis vector in the $x$-direction. Let us check that the electric field is indeed given by the gradient of $V$. First, the fact that $V(\boldsymbol{x})$ is independent of $y$ and $z$ implies that

$$\frac{\partial V}{\partial y} = \frac{\partial V}{\partial z} = 0.$$

Thus, the $y$ and $z$ components of the electric field are zero. Furthermore, one has

$$\frac{\partial V}{\partial x} = -\frac{Q}{4\pi\epsilon_0}\frac{x}{(x^2+a^2)^{3/2}},$$

where we have used the chain rule. Then

$$\boldsymbol{E} = -\nabla V = \frac{Q}{4\pi\epsilon_0}\frac{x}{(x^2+a^2)^{3/2}}\hat{\boldsymbol{\imath}},$$

in agreement with the previous result.

## Exercises

(1) Point charges $Q_1$ and $Q_2$ are placed on the $y$-axis, as shown in Figure 3.23. Show that the electric field at point $P$ on the $x$-axis is given by

$$\boldsymbol{E} = \frac{1}{4\pi\epsilon_0(x^2+L^2)^{3/2}}[x(Q_1+Q_2)\hat{\boldsymbol{\imath}} + L(Q_2-Q_1)\hat{\boldsymbol{\jmath}}],$$

where $\hat{\boldsymbol{\imath}}$ and $\hat{\boldsymbol{\jmath}}$ are unit vectors in the $x$ and $y$ directions, respectively.

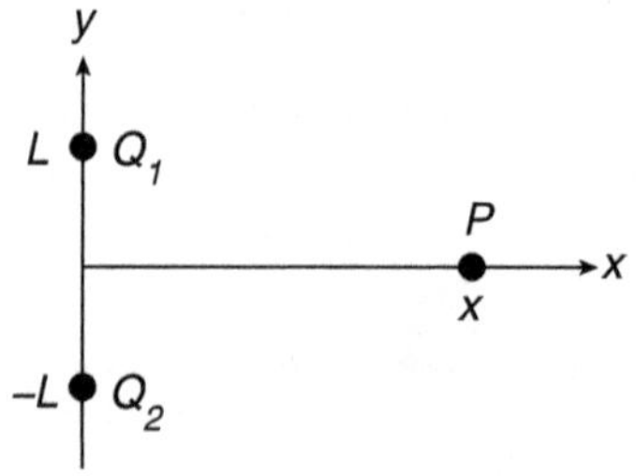


Fig. 3.23 Two charges located along the $y$ axis.

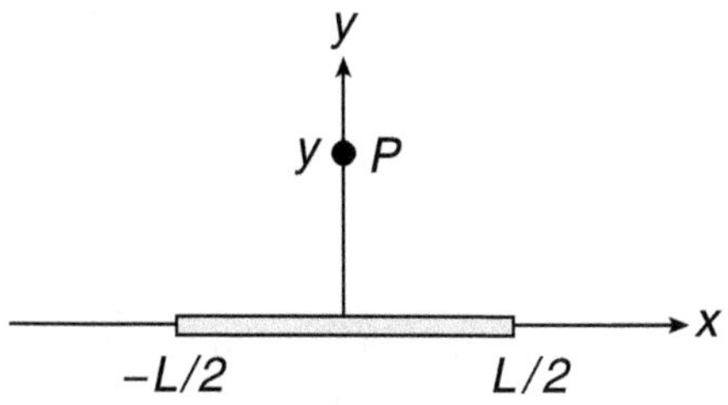


Fig. 3.24 A charged rod.

(2) A dipole is placed at $P$ in Figure 3.23, with dipole moment

$$\boldsymbol{\mu} = \mu \hat{\boldsymbol{\imath}}.$$

Assuming that the separation of charges in the dipole is small compared to $x$ and $L$, show that the torque on the dipole is given by

$$\boldsymbol{G} = \frac{\mu L(Q_2 - Q_1)}{4\pi\epsilon_0(x^2 + L^2)^{3/2}} \hat{\boldsymbol{k}}.$$

How does this become more complicated if the separation in the dipole is *not* small compared to $x$ and $L$?

(3) A uniformly charged rod of length $L$ carries a total charge $Q$, and is oriented along the $x$-axis as shown in Figure 3.24. Show that the field at a point $P$ at distance $y$ along the $y$-axis is given by

$$\boldsymbol{E} = \frac{Q}{2\pi\epsilon_0 y(L^2 + 4y^2)^{1/2}} \hat{\boldsymbol{\jmath}}.$$

(4) An electron is at the centre of a cube. What is the electric flux through each face of the cube? Is this positive or negative, and why?

(5) Consider a spherically symmetric charge distribution confined within a sphere of radius $R$. Show that, outside the sphere, the electric field behaves as if all of the charge is concentrated at the centre of the sphere. What goes wrong in your calculation if the charge distribution is *not* spherically symmetric?

(6) Consider the charge density (in three spatial dimensions)

$$\rho(r) = \begin{cases} \lambda r^2, & r \leq R \\ 0, & r > R, \end{cases}$$

where $r$ is the radial distance from the origin. Show that the magnitude of the electric field is given by

$$E = \frac{\lambda r^3}{5\epsilon_0},$$

for $r \leq R$. What about for $r > R$?

(7) The potential difference on moving a charge $q$ from $\infty$ to a point $\boldsymbol{x} = (x, y, z)$ is given, for a certain charge configuration, by

$$V = \alpha \frac{x + y - L}{x^2 + y^2 + L^2},$$

for some constant $\alpha$. What is the electric field?

# Chapter 4

# A First Look at Circuits

In the previous chapter, we have started to work towards a general understanding of electricity in terms of the *electric field*, which will play a crucial role in the complete theory of electromagnetism. Before continuing our development of this theory, however, it is worth examining some further applications of electricity. We will also introduce some concepts that we will use heavily later on, beginning in the following section.

## 4.1 Capacitance

As we have seen previously, conductors can store excess charge on their surfaces. Indeed, the charge must reside on the surface, as the electric field is zero inside a conductor, which implies zero net charge inside by Gauss' law. The ability of a given system of conductors to store charge can be measured by defining the *capacitance*. The precise definition of this quantity depends on whether we are talking about a single conductor, or a pair of conductors.

A single conductor has *self-capacitance*. In defining this, we can first note that the potential $V$ is equal everywhere on the surface of a conductor. If it wasn't, there would be a component of the electric field $\boldsymbol{E} = -\nabla V$ parallel to the surface, and thus the charges on the surface would move about in order to cancel it. Given that $\boldsymbol{E}$ and $V$ are both linear functions of charge, it must be true that the total charge carried by the conductor obeys

$$Q \propto V.$$

The self-capacitance is then defined to be the constant of proportionality $C$, where

$$Q = CV. \tag{4.1}$$

A higher $C$ means that the conductor can store more charge for a given potential, and thus in a sense $C$ measures the capacity to store charge, hence the name "capacitance". The SI unit of capacitance is Coulombs per volt, also referred to as the Farad: $1\ \mathrm{F} \equiv 1\mathrm{CV}^{-1}$.

As an example of self-capacitance, consider a conducting sphere of radius $R$, with a total excess charge $Q$ distributed on its surface (this charge must be uniformly distributed, so that no electric field parallel to the surface occurs at any point). To find the self-capacitance $C$, we must first find the potential on the sphere, i.e. the potential difference upon bringing a test charge from infinity to the surface of the sphere. This will be given by

$$V = -\int \boldsymbol{E} \cdot d\boldsymbol{l},$$

where for convenience we can choose a radial path from $r = \infty$ to $r = R$. The electric field can be found from Gauss' law, using a spherical surface of radius $r > R$, as shown in Figure 4.1. By symmetry, the field is pointing in the radial direction, and is constant on the spherical surface. Gauss' law then gives

$$|\boldsymbol{E}| 4\pi r^2 = \frac{Q}{\epsilon_0},$$

so that

$$\boldsymbol{E} = \frac{Q}{4\pi\epsilon_0 r^2}\hat{\boldsymbol{r}},$$

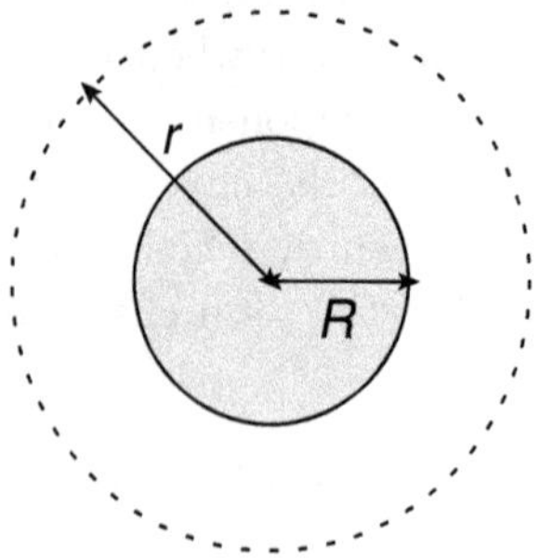


Fig. 4.1 A spherical surface of radius $r$ outside a conductor of radius $R$.

where $\hat{\boldsymbol{r}}$ is a unit vector in the radial direction. The potential on the surface of the sphere (i.e. at $r = R$) is then given by

$$V(R) = -\int_{\infty}^{R} dr \frac{Q}{4\pi\epsilon_0 r^2} = \frac{Q}{4\pi\epsilon_0 R}.$$

Applying Eq. (4.1), we find that the self-capacitance is given by

$$C = 4\pi\epsilon_0 R.$$

We see that $C$ is indeed constant for a given surface, and depends on its geometry (here the value of $R$), and the permittivity of the space (here $\epsilon_0$, as the sphere is in a vacuum).

It is actually quite rare to see self-capacitance being used. A much more common definition is that of the *mutual capacitance* (often just called the *capacitance*) of a pair of conductors. If the latter carry equal and opposite charges, we may define the mutual capacitance via

$$Q = C\Delta V, \tag{4.2}$$

where $\Delta V$ is the potential difference between the positively charged and negatively charged plates. More precisely,

$$\Delta V = V_+ - V_-,$$

where $V_\pm$ is the potential on the positive/negative plate, respectively. Thus, $\Delta V$ is chosen to be positive, so that the mutual capacitance $C$ turns out to be a positive number. Such pairs of conductors are said to form a *capacitor*, and these devices have many applications in electrical circuits. As an example, let us consider a *parallel plate capacitor*, consisting of two charged planes as in Figure 3.22, which can usually be approximated by infinite planes when calculating the field: the field will deviate from that of infinite planes only at the edges of the capacitor, and such edge effects are usually not significant provided the width of the planes is much greater than their separation $L$. The capacitance is defined by Eq. (4.2), and we must then find the potential difference between the plates. Earlier, we saw that we can choose the potential on the upper plate in Figure 3.22 to be $V_+ = 0$, such that the potential on the lower plate is then

$$V_- = -\frac{\sigma L}{\epsilon_0}.$$

This then implies

$$\Delta V = \frac{\sigma L}{\epsilon_0} = \frac{QL}{\epsilon_0 A},$$

where on the right we have used $\sigma = Q/A$, where $Q$ is the total charge on the positive plate, and $A$ its area. Using Eq. (4.2), we then get

$$C = \frac{Q}{\Delta V} = \frac{\epsilon_0 A}{L}.$$

As for self-capacitance, we find that the mutual capacitance depends on the geometry of the system (the area of the plates $A$, and their mutual separation $L$), and also the permittivity of space $\epsilon_0$. Furthermore, it makes sense that $C$ is higher for larger $A$: it measures the ability of the system of conductors to store charge, and they can clearly store more charge if the area of the plates is greater.

## 4.2 Dielectrics

The capacitors we considered in the previous section were in a vacuum. However, usually capacitors have some insulating material in between the two conducting surfaces, as shown in Figure 4.2. The basic idea of this is that one wants a capacitor to store charge, and it can store more charge provided we make it harder for charge to be transferred from one charged surface to the other. Such materials are usually called *dielectrics*, and they indeed affect the capacitance. Experiment shows that the presence of a dielectric *lowers* the potential difference between the conductors, for a given charge $Q$. Let the potential difference between the plates in Figure 4.2

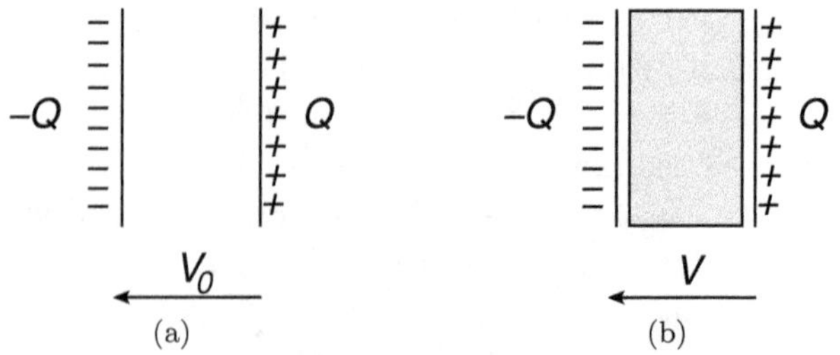


Fig. 4.2 (a) A parallel plate capacitor in a vacuum, with potential difference $V_0$ between the plates; (b) a dielectric lowers the potential difference to $V < V_0$.

be $V_0$. Then, the capacitance of the vacuum-filled capacitor is

$$C_0 = \frac{Q}{V_0}.$$

If the potential of the dielectric-filled capacitor in Figure 4.2 is $V < V_0$, the capacitance is

$$C = \frac{Q}{V} > C_0.$$

Thus, a dielectric *increases* the capacitance (the ability to store charge per unit potential difference). For a given material, we can define the *dielectric constant*

$$K = \frac{C}{C_0}, \tag{4.3}$$

namely the capacitance of a capacitor filled with the material, divided by the capacitance if the dielectric is absent. This is a dimensionless number, and furthermore does not depend on the geometry of a particular capacitor: any geometric factors (such as radii, areas or lengths) must cancel out to leave a dimensionless quantity. Some examples of the dielectric constants of different materials are shown in Table 4.1. Recall from earlier that for a parallel plate capacitor in a vacuum, one has

$$C_0 = \frac{\epsilon_0 A}{L},$$

where $A$ is the area of each plate, and $L$ the mutual separation between them. Likewise, for concentric spheres, one has (see the exercises)

$$C_0 = \frac{4\pi\epsilon_0 R_1 R_2}{R_2 - R_1},$$

Table 4.1 The dielectric constant of different materials, where air is taken to be at 1 atmosphere pressure.

| Material | $K$ |
|---|---|
| Vacuum | 1 (by definition) |
| Air | 1.00059 |
| Mylar | 3.1 |
| Glass | 5–10. |

Note that Mylar is a type of plastic used in commercial capacitors.

where $R_1$ and $R_2$ are the radii of the inner and outer spheres, respectively. In both cases, $C_0 \propto \epsilon_0$, a fact that ultimately comes from Gauss' law: the capacitance is related to the inverse of the potential, where the latter is linear in the electric field. The electric field in turn goes like the inverse of the permittivity from Gauss' law, so that the capacitance has a positive power of $\epsilon_0$. The relation $C = KC_0$ then means that

$$C \propto K\epsilon_0$$

in general. That is, the presence of the dielectric modifies the permittivity according to

$$\epsilon_0 \to \epsilon = K\epsilon_0.$$

This gives an alternative definition of $K$, as the *relative permittivity* of a material:

$$K = \frac{\epsilon}{\epsilon_0}. \tag{4.4}$$

That is, the dielectric constant $K$ is the ratio of the permittivity of a material, to the permittivity of free space. The relative permittivity is also denoted by $\epsilon_r$, and you may be wondering why there are two symbols for the same thing! The answer is that we have here considered *static* fields, and *isotropic* materials (i.e. materials with no preferred direction). Permittivity is more complicated otherwise, such that the dielectric constant and the relative permittivity can indeed be different things.

It is natural to ask what causes the potential difference to change if a dielectric is present. This is due to *polarisation* of the material. A given dielectric will be electrically neutral overall, but will be made of e.g. molecules, which contain charged particles. In each molecule, there may be a net charge at each end, as shown schematically in Figure 4.3(a). For an isotropic dielectric material, the dipoles are typically randomly ordered (e.g. they will have thermal energy, which makes them rotate and have no specific orientation), shown schematically in Figure 4.3(b). However, we saw earlier in Section 3.5 that a dipole in an electric field experiences a torque

$$\boldsymbol{G} = \mu \times \boldsymbol{E},$$

where $\mu$ is the dipole moment. The effect of this is to make the dipole align with the electric field, and it follows that if we apply an electric field to a dielectric, the molecular dipoles will tend to line up as in Figure 4.3(c).

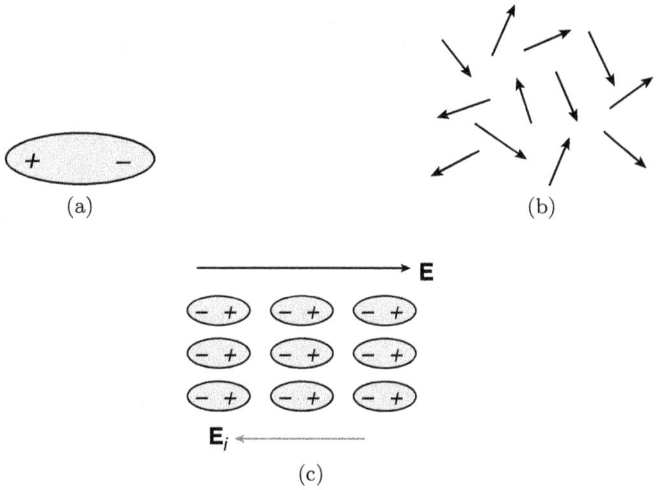


Fig. 4.3 (a) A neutral molecule can have an imbalance of charge, that looks like an electric dipole; (b) in the absence of an applied field, the orientation of molecular dipoles in a dielectric is random, and thus there is no net polarisation; (c) if a field is applied, the dipoles line up with each other, creating a net polarisation.

This creates a net electric field in the material, that *opposes* the applied field. We call this the *induced* electric field $\boldsymbol{E}_i$, and depict it by the lower arrow in Figure 4.3. By the principle of superposition, the net field inside the dielectric is given by

$$\boldsymbol{E}_{\text{net}} = \boldsymbol{E} + \boldsymbol{E}_i = (E - E_i)\hat{\boldsymbol{\imath}},$$

where we have taken the applied field to be in the $x$-direction, and defined $E = |\boldsymbol{E}|$, etc. Typically, $E_i < E$ in general, thus the induced field does not cancel out the applied field completely. However, it means that the field inside the dielectric is *smaller* in magnitude than it would be in a vacuum.

The above situation is what happens if a dielectric is put between the plates of a capacitor. If $\boldsymbol{E}_0$ is the field that would be present between the plates in a vacuum, then the field — if the dielectric is present — has magnitude $E_0 - E_i$, such that the potential difference between the plates becomes

$$\Delta V = (E_0 - E_i)L,$$

which is indeed *less* than the vacuum case.

Note that in the above discussion, we assumed that dipoles already existed in the dielectric material, before the field was applied. This is not

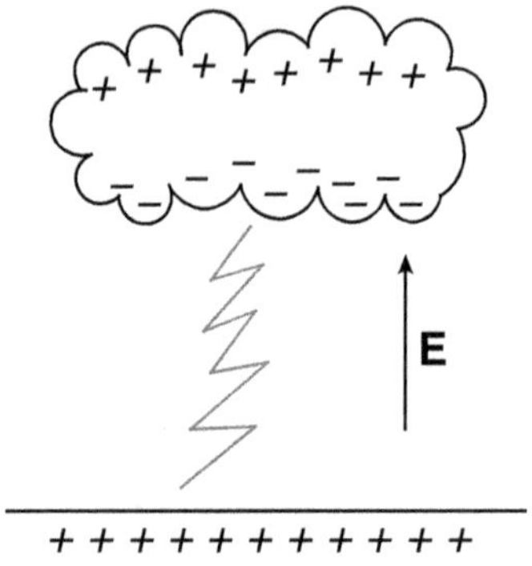


Fig. 4.4 Lightning: dielectric breakdown of air.

necessarily the case: the material may be made of molecules which have no charge imbalance. However, even in this case the presence of an applied electric field can induce a charge imbalance in the molecules, such that the mechanism of polarisation is very similar. We also assumed that the induced electric field $\boldsymbol{E}_i$ is simply proportional to the applied field. This is indeed an assumption, but it turns out to be an approximation that works well for sufficiently weak applied fields. If the applied field instead becomes very strong, it can strip electrons away from molecules, such that these become mobile and the material becomes conducting rather than insulating. This is known as *dielectric breakdown*, and you will be familiar with a dramatic example from everyday life: lightning, which corresponds to dielectric breakdown of air! In a storm, we know that an imbalance of charge builds up in clouds, although the mechanism for this to happen is still not fully understood. The net charge on the lower cloud surface repels charged particles in the ground, leading to a net charge there too, whose sign is opposite to that on the bottom of the cloud. This creates an electric field between the cloud and the ground, and when this becomes large enough (due to sufficient charge building up in the cloud), dielectric breakdown occurs, and lightning strikes!

## 4.3 Electric Current

So far we have looked at static situations only. In general, however, charged particles can move, and create a flow of charge. A flow of charge is more properly called a *current*, and it is clearly important to understand currents if we are to fully understand electricity. By definition, only conducting materials can carry currents, as it is only conductors that have charged particles that are free to move. This is not all we need for a current to flow: we also need to apply an electric field across the material. To see why, consider a silly theorist's picture of what happens in a metal, as we have already seen in Figure 3.1. This consists of a crystal lattice of ions, surrounded by relatively mobile electrons. At room temperature, these electrons will be moving with a very high velocity, due to what is called their *thermal motion*. However, they will not get very far through the metal, as they will repeatedly collide with the ions, and bounce off (i.e. they will be repelled by other electrons in the ions, even though the ions have a net positive charge). Furthermore, the direction of each electron velocity is essentially random, so that there is no *net* flow of electrons to the left or to the right.

Now consider applying an electric field acting, say, to the right. This will cause each electron to feel a force

$$\boldsymbol{F} = -e\boldsymbol{E}$$

to the left (electrons are negatively charged). However, each electron still keeps colliding with ions, so that the overall motion is that of very fast moving electrons repeatedly colliding with the ions in the metal, together with a relatively slow overall drift of the electrons to the left. It is the latter that constitutes a flow of current, and we say that the electrons have a *drift velocity* $\boldsymbol{v}_d$ on average. In the metal example we have considered, the mobile charged particles are electrons, and thus there is a net flow of negative charge to the left. This is equivalent to a net flow of *positive charge* to the right. By an annoying historical convention (n.b. current was known about before the existence of electrons), the direction of electric current is defined to be the direction of flow of *positive charge*. Thus, in our metal example, there is a current flowing to the right, even though the electrons are actually moving to the left.

Let us now more formally define the current. Consider a conducting wire with cross-section area $A$, as shown in Figure 4.5, and where an electric field $\boldsymbol{E}$ is applied to the right. Any positive mobile charges will move to the right due to the electric field. Similarly, any mobile negative charges will move to the left. In both cases, this results in a net flow of *positive* charge to the *right*, and thus a current to the right. That is, the current is always in the direction of the applied electric field. Imagine that a positive charge $dQ$ crosses the cross-sectional surface $S$ in time $dt$. Then the current is defined as

$$I = \frac{dQ}{dt}, \tag{4.5}$$

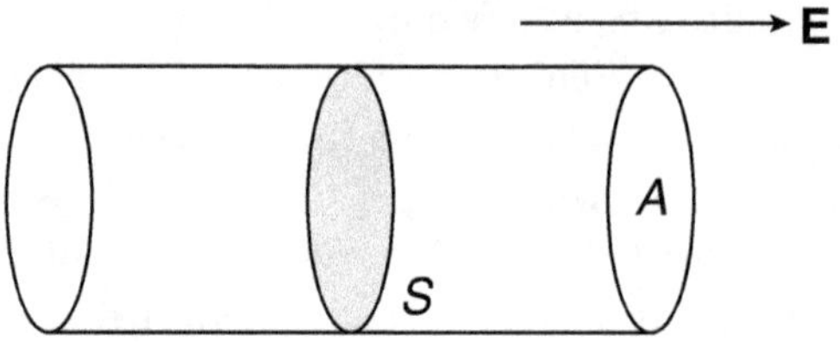


Fig. 4.5 A conducting wire in an applied electric field $\boldsymbol{E}$, where a cross-sectional surface $S$ is considered.

i.e. it is the rate of flow of positive charge in the direction of $\boldsymbol{E}$. The SI unit of current is the Ampere A, where $1\text{A} \equiv 1\,\text{Cs}^{-1}$ (Coulomb per second).

We can also relate the current to the drift velocity. Imagine that there are $n$ mobile charged particles per unit volume, each of which has charge $q$. In time $dt$, these will travel a distance $|\boldsymbol{v}_d|dt$, and thus trace out a volume

$$dV = A|\boldsymbol{v}_d|dt.$$

Given each carrier has charge $q$, there is a charge $nq$ per unit volume. Thus, the total positive charge crossing $S$ in the direction of $\boldsymbol{E}$ is then

$$dQ = n|q|dV = n|q|A|\boldsymbol{v}_d|dt.$$

Care is needed here with signs. The charge $q$ can be positive or negative. However, the flow of positive charge is always in the direction of $\boldsymbol{E}$ (as we saw above). Thus, putting $|q|$ rather than $q$ in the above result guarantees that this is always positive, whether $q < 0$ or $q > 0$, as required. Finally, we arrive at an expression for the current:

$$I = \frac{dQ}{dt} = nA|q||\boldsymbol{v}_d|. \tag{4.6}$$

Given that wires can have different geometries, another useful quantity to define is the *current density*

$$J = \frac{I}{A},$$

namely the current (rate of flow of positive charge in the direction of $\boldsymbol{E}$) per unit cross-sectional area. From above, we find that

$$J = n|q||\boldsymbol{v}_d|.$$

Note that $I$ and $J$ are both scalar quantities, albeit with a sign: current in the wire is flowing to the right or to the left. One can also define a *vector current density*

$$\boldsymbol{J} = nq\boldsymbol{v}_d, \tag{4.7}$$

whose magnitude is equal to the scalar current density $J$, and whose direction is that of the drift velocity. The SI units of current density are $\mathrm{C\,s^{-1}\,m^{-2}}$ (Coulombs per second per metre squared). Note that, although we have considered the particular case of current in a wire, the idea of a vector current density is quite general. Consider an arbitrary continuous charge distribution with density $\rho(\boldsymbol{x})$, such that the velocity of the charge at some point $\boldsymbol{x}$ is $\boldsymbol{v}$. The current density at $\boldsymbol{x}$ is then given by

$$\boldsymbol{J} = \rho(\boldsymbol{x})\boldsymbol{v},$$

which follows from similar arguments as those leading to Eq. (4.7).

## 4.4 Resistivity and Resistance

We have seen that we need a conducting material, together with an electric field applied across this material, to cause a current to flow. The current density $J$ depends on the drift velocity $\boldsymbol{v}_d$, which itself depends upon how easily mobile charged particles can move through the material. This is clearly a very complicated process in general, and cannot be described exactly due to the tremendous number of particles involved. In many cases, however, the induced current density turns out to be simply proportional to the applied field. That is, we can write $E = \rho J$ in general, where $E$ and $J$ are the magnitudes of the electric field and current density, respectively. This coefficient of proportionality is called the *resistivity*

$$\rho = \frac{E}{J}. \tag{4.8}$$

A perfect conductor will have $\rho = 0$, and a perfect insulator will have $\rho = \infty$. That is, Eq. (4.8) implies that if $\rho$ is *higher*, the induced current density is *less* for a given applied field $E$. In practice, the resistivity $\rho$ is not a constant, but depends on, e.g. the temperature of the material. In our metal example, we have seen that multiple collisions of electrons with the ions in the material mean that it is difficult to get much of a drift velocity, where the latter constitutes the flow of current. If the metal is hotter, the electrons have more thermal energy, and thus collide more with the ions.

One thus expects that the resistivity would *increase* with temperature.[1] In the literature, you may also see people talking about *conductivity* $\sigma$ instead of resistivity $\rho$. The relationship between these quantities is simply

$$\sigma = \frac{1}{\rho}, \tag{4.9}$$

so that one has

$$J = \sigma E. \tag{4.10}$$

It follows that a perfect conductor has $\sigma = \infty$, and a perfect insulator has $\sigma = 0$.

Now consider a length of conducting material $L$ with cross-sectional area $A$, carrying current $I$. For a constant applied field, the potential difference across the material will be

$$\Delta V = -EL.$$

From the definitions of resistivity and current density, we also have that

$$E = \rho J = \frac{\rho I}{A}. \tag{4.11}$$

Combining the two expressions for the electric field, we thus find

$$|\Delta V| = \left(\frac{\rho L}{A}\right) I,$$

from which it follows that if the resistivity $\rho$ is constant, the potential difference across the material is proportional to the current. The constant of proportionality is usually called the *resistance*

$$R = \frac{\rho L}{A}, \tag{4.12}$$

and a material with non-zero resistance is called a *resistor*. Often people simply omit the $\Delta$ symbol in the above equation, and also the magnitude signs, and write it as

$$V = IR. \tag{4.13}$$

[1] There are other materials (e.g. semiconductors, superconductors) where the relationship of resistivity with temperature is more complicated, but we will not consider these in this book.

That is, the size of the potential difference across a resistor is equal to the resistance multiplied by the current. This is *Ohm's law*, and we see that it is not an exact law of nature! Rather, it is an approximation, based ultimately on the fact that the induced current density is proportional to the applied electric field in simple materials. Conducting materials that do not obey Ohm's law are called *non-ohmic conductors*. People usually measure resistance in *Ohms* $\Omega$, where $1\Omega \equiv 1\,\mathrm{VA}^{-1}$ (volts per amp). This in turn implies that resistivity $\rho$ has units $\Omega$m (Ohm metres).

## 4.5 Electromotive Force and Circuits

Isolated conductors will not carry a current for very long. Consider, for example, applying an electric field towards the right of the length of conducting material in Figure 4.5. This causes a current, such that electrons move to the left in the wire. This in turn leads to a build up of negative charge on the left-hand surface of the conductor, and a net positive charge on the right-hand surface. There is then an *induced electric field* $\boldsymbol{E}_i$ between these charges, and when enough charge has built up, the induced field precisely cancels the applied field, so that the net electric field inside the conductor is zero. No current then flows!

We can instead maintain a current by connecting conductors in closed loops, called *electrical circuits*. But we then still need something extra to get the current to flow. Earlier we saw that the change in potential $V$ around any closed loop in space is zero (this follows from the conservative nature of the electric field). However, we saw in the previous section that the potential *decreases* across a resistor, in accordance with Ohm's law. Thus, for a current to flow around a closed loop with any resistance (which will always be there for real materials), there must be something that creates positive potential differences, so that $\Delta V = 0$ around the closed loop. This something is called, for historical reasons, *electromotive force*, or EMF for short. Sources of EMF include batteries, generators, etc., and we will not explore how these devices work in this book. All that matters for our purposes is that any given source has a *positive* and a *negative terminal*, such that the change in potential from the negative to the positive terminal is

$$\mathcal{E} = \Delta V > 0. \tag{4.14}$$

Despite the name, EMF is not a force! It is just the potential difference across the battery, generator, or similar device. Once we have an EMF source, we can make circuits by joining together various components to

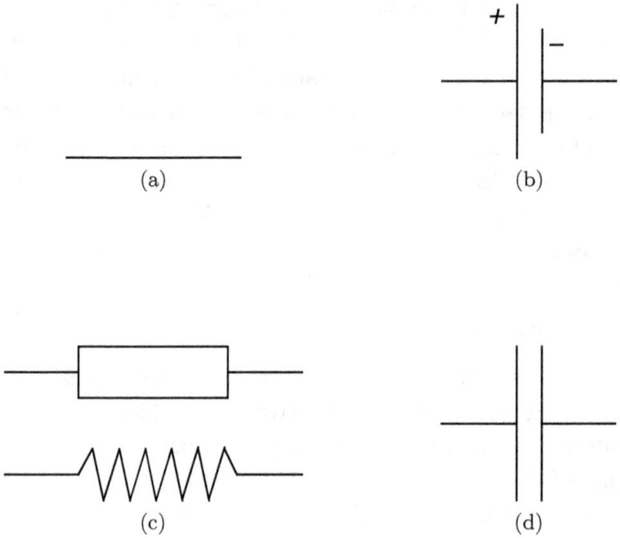


Fig. 4.6 Conventional symbols for various electrical components: (a) an ideal wire (no resistance); (b) a source of constant electromotive force (EMF); (c) a resistor; (d) a capacitor.

make one or more loops. To draw these circuits, it is convenient to use symbols that have become established convention for different electrical components. We show a selection of these in Figure 4.6.

Perhaps the simplest non-trivial circuit we can think of has an EMF source and a single resistor, as shown in Figure 4.7(a). Using Ohm's law, the total change in potential difference $V$ around this loop is

$$\mathcal{E} - IR = 0, \Rightarrow \mathcal{E} = IR,$$

i.e. we can relate the EMF of the source to the current flowing, and the nature of the components in the circuit (in this case the resistance). Real sources of EMF are not usually ideal, as charges can experience a resistance upon passing through the device. In simple cases, this can be modelled by including a small *internal resistance* $r$ in the circuit, as shown in Figure 4.7(b).

More complicated circuits have more components and/or loops, and we often want to find currents or potential differences in various different parts of a circuit. To this end, we can use *Kirchhoff's laws.* There are two of these. One relates to current and the other to voltage, or potential

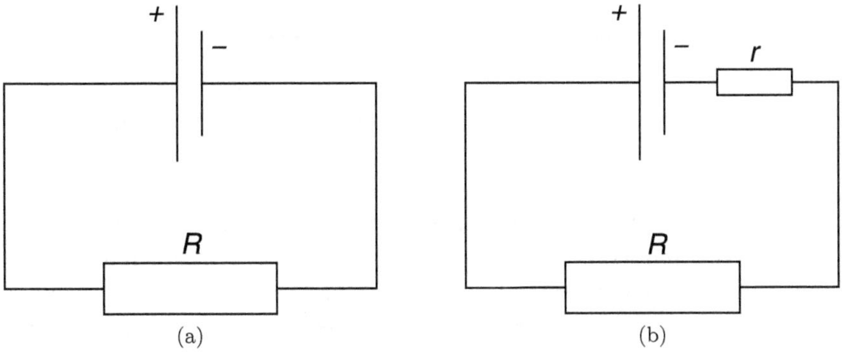


Fig. 4.7 (a) Circuit consisting of an EMF source and a single resistor; (b) an added internal resistance for the EMF source.

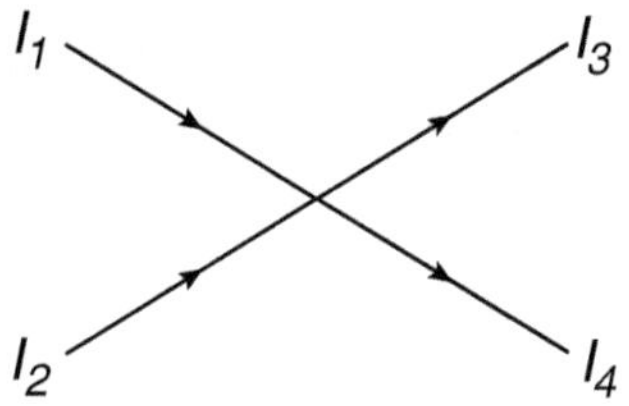


Fig. 4.8 Currents entering and leaving a junction.

difference. However, they are really just consequences of underlying physics that you have already seen:

(1) *Kirchhoff's voltage law*: This states that the sum of all EMFs and potential differences around any closed loop of a circuit is zero. As stated above, this follows directly from the conservative nature of the electric field.
(2) *Kirchhoff's current law*: This states that the sum of all currents entering a junction equals the sum of all currents leaving a junction. As an example, consider Figure 4.8, which shows two currents $I_1$ and $I_2$ flowing in, and currents $I_3$ and $I_4$ flowing out. Kirchhoff's current law in this case is

$$I_1 + I_2 = I_3 + I_4.$$

The law follows simply from the fact that charge cannot be created or destroyed (i.e. charge is conserved). Thus, all charge that flows into a junction must flow out of it somewhere.

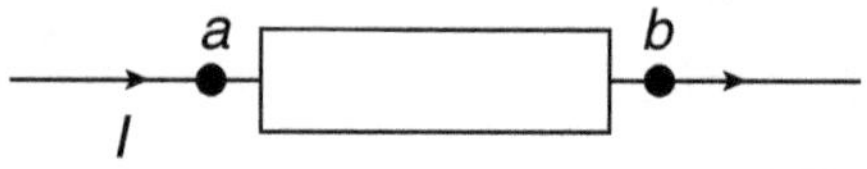


Fig. 4.9 A resistor, with different points labelled. We may then choose to define the potential difference in either direction, relative to the direction of the current $I$.

Before we illustrate the use of these laws, some comments are in order regarding different conventions for talking about potential differences. Formally, the potential difference between two points can be obtained as minus the line integral of the electric field, as defined in Eq. (3.28). As we have seen above, this results in a negative potential difference if we cross a resistor in the direction of the current. If we wanted to, however, we could define the potential difference the other way around, which would give us a positive number. This can be useful, only in that it avoids a proliferation of minus signs in various equations. Following the notation of Eq. (4.13), let us use a single letter $V$ to denote the potential difference across a resistor (or other component) in the opposite direction to the current. Referring to Figure 4.9, we thus have, by definition,

$$\Delta V = V_b - V_a, \quad V = -\Delta V = V_a - V_b, \tag{4.15}$$

where $V_i$ is the potential at point $i$.

One consequence of Kirchhoff's laws is that we can simplify circuits containing multiple resistors and/or capacitors. Consider first the case of *resistors in series*, as shown in Figure 4.10(a). From Ohm's law applied to each resistance, and using Eq. (4.15), we have

$$V_1 = IR_1, \quad V_2 = IR_2.$$

Thus, the total potential difference is

$$V_{\text{tot.}} = V_1 + V_2 = I(R_1 + R_2).$$

We therefore see that the two resistors look like a *single* effective resistance

$$R = R_1 + R_2. \tag{4.16}$$

We can also consider resistors in parallel, as shown in Figure 4.10(b). In this case, Kirchhoff's current law applied to the junction on the left implies

$$I = I_1 + I_2.$$

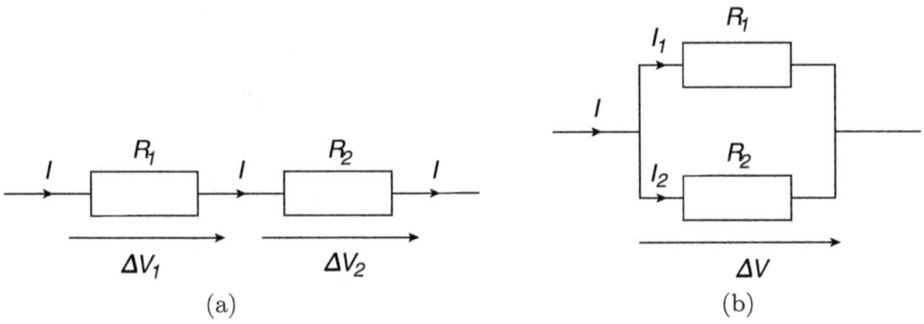


Fig. 4.10 (a) Two resistors in series; (b) two resistors in parallel.

Furthermore, the voltage law applied to the loop implies that the potential difference across each resistor must be the same, as labelled in the figure. The two resistors again look like a single effective resistance $R$, which by definition will be given by

$$V = IR.$$

Using the above result for the currents, we can then write

$$\begin{aligned}\frac{1}{R} &= \frac{I}{V}\\ &= \frac{I_1 + I_2}{V}\\ &= \frac{I_1}{V} + \frac{I_2}{V}.\end{aligned}$$

Applying Ohm's law to each resistor separately, we then arrive at

$$\frac{1}{R} = \frac{1}{R_1} + \frac{1}{R_2}. \tag{4.17}$$

Similar formulae can be derived for capacitors. Consider first the case of two capacitors in series, as shown in Figure 4.11(a), and where capacitor $C_i$ has charge $Q_i > 0$ on its positive (left-hand) plate, and $-Q_i$ on its negative (right-hand) plate. If the potential difference across capacitor $C_i$ is $\Delta V_i$, the definition of capacitance implies

$$Q_i = -C_i \Delta V_i. \tag{4.18}$$

Care is needed here to get the sign right. Recall that the potential difference entering the definition of the capacitance is

$$V_+ - V_-,$$

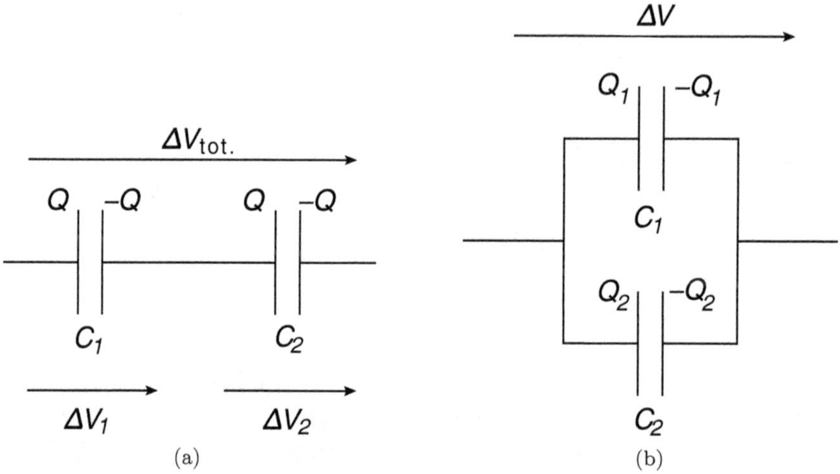


Fig. 4.11 (a) Two capacitors in series; (b) two capacitors in parallel.

if $Q > 0$, where $V_{\pm}$ is the potential on the positive/negative plate, respectively. The positive plates of each capacitor are on the left in Figure 4.11, given that current (the flow of positive charge) is flowing to the right. Thus, the potential difference $\Delta V$ as measured by integrating the electric field around the circuit is defined in the *opposite* direction to the potential difference entering $C$. It is perhaps more straightforward to adopt the second convention in Eq. (4.15), and thus to rewrite Eq. (4.18) as

$$Q_i = C_i V_i, \tag{4.19}$$

where $V = -\Delta V$ is the potential difference measured in the opposite direction to the current. We are being very finicky here, but the main thing to remember is that, as for resistors, the potential *drops* as we go across a capacitor.

Returning to the situation of Figure 4.11, given that charge is not created or destroyed in the middle wire, we must have that

$$Q_1 = Q_2 \equiv Q,$$

as shown explicitly in the figure. Then, the pair of capacitors looks like a *single* capacitor, with effective capacitance given (by definition) by

$$Q = CV_{\text{tot.}},$$

where $V_{\text{tot.}}$ is the total potential difference across both capacitors. Given that

$$V_{\text{tot.}} = V_1 + V_2,$$

this implies

$$\frac{Q}{C} = \frac{Q}{C_1} + \frac{Q}{C_2},$$

and thus

$$\frac{1}{C} = \frac{1}{C_1} + \frac{1}{C_2}. \tag{4.20}$$

Now consider capacitors in parallel, as in Figure 4.11(b). By Kirchhoff's voltage law, the potential difference across each capacitor is the same, and the definition of capacitance then gives

$$Q_1 = C_1 V, \quad Q_2 = C_2 V.$$

In this case, $Q_1 \neq Q_2$ in general, but it is nevertheless true that we can regard the system as a single capacitor, with a capacitance $C$ defined by

$$Q_{\text{tot.}} = CV,$$

where

$$Q_{\text{tot.}} = Q_1 + Q_2$$

is the total charge on the positive plates of the two capacitors. Combining the above equations yields

$$CV = C_1 V + C_2 V,$$

and thus

$$C = C_1 + C_2. \tag{4.21}$$

Thus, for capacitors in parallel, we add the capacitances. For capacitances in series, the reciprocal of the overall capacitance is given by the sum of the reciprocals of the individual capacitances. This is the other way round to when we combine resistances.

You may be wondering what the above results are good for. One very practical use is if you are building a circuit, and you need a certain resistance or capacitance, but haven't managed to buy one of these from a shop. You can then combine the resistors or capacitors that you *do* have lying around, according to the above rules!

## 4.6 Energy and Power in Circuits

The presence of components in electrical circuits means that energy can be lost from a circuit, where this energy is supplied by the EMF source. Often this is a useful thing, such as when electrical energy is turned into motion (a motor), light (a bulb), heat (a kettle), or sound (a loudspeaker). For a given component with potential difference $\Delta V$ across it, we can derive an expression for its *power* (i.e. energy gained/lost per unit time). Consider a charge $q$ passing through the component — by definition, this will experience a change in electrical potential energy

$$\Delta U = q\Delta V$$

(recall that the potential difference was defined as the change in potential energy per unit charge). If $\Delta V < 0$ (e.g. across a resistor), potential energy is lost. It may be converted to heat, or different forms of energy. If $\Delta V > 0$ (e.g. a battery), the component does work on the charge. A battery, for example, converts chemical energy into the potential energy of the charged particles at the positive terminal.

For a given component, let a charge $dq$ pass through in time $dt$. Then the change in potential energy is

$$dU = dq\Delta V = \frac{dq}{dt}\Delta V dt = I\Delta V dt,$$

where we have recognised $I = dq/dt$ as the definition of the current through the component. The power of the component is the change in energy per unit time:

$$P = \frac{dU}{dt},$$

and thus

$$P = I\Delta V \tag{4.22}$$

is the power of an electrical component, in terms of the current flowing through it, and the potential difference across it. As an example, a resistor has

$$\Delta V = -IR,$$

It thus *dissipates* energy at a rate

$$|P| = I^2 R = \frac{(\Delta V)^2}{R}. \tag{4.23}$$

Typically, this is lost as heat energy. We also noted above that the potential drops across a capacitor, which would again result in energy apparently being lost from the circuit. In fact, it gets stored in the electric field inside the capacitor, given that this carries an energy density as given by Eq. (3.33). Unlike the case of a resistor, this energy can be recovered if we wish by e.g. discharging the capacitor at a convenient time.

In this and the previous chapter, we have dealt with electricity in isolation. Importantly, we have introduced the concept of the *electric field*, and argued that this is the central quantity for describing electric behaviour, that has a real life of its own in terms of being able to carry energy, momentum, etc. The cases we have looked at have all been very special, however, involving charges that are not moving, or currents that are not changing with time. This is clearly only a subset of things that can actually happen in nature, so that we need to generalise further if we are to fully understand electricity. What's more, it turns out that when we let charges move, things immediately get a lot more complicated, and we can no longer consider electric behaviour alone. Moving charges unavoidably lead to *magnetic* behaviour, which we will start to address in the following chapter.

## Exercises

(1) Consider a sphere of radius $R_2$ holding charge $Q$, and another concentric sphere of radius $R_1 < R_2$ holding charge $-Q$. Show that the capacitance of this system is given by

$$C = \frac{4\pi\epsilon_0 R_1 R_2}{R_2 - R_1}.$$

(2) A copper wire with cross-sectional area $1 \times 10^{-6}$ m$^2$ carries a current of 6 A.

(i) What is the electric field in the wire? The resistivity of copper is $1.68 \times 10^{-8}\Omega$m.

(ii) If the number of free electrons per unit volume is $8.5 \times 10^{28}\,\text{m}^{-3}$, what is the magnitude of the drift velocity of the electrons? How does this compare to a typical human walking speed, and does this result surprise you?

# Chapter 5

# Introducing Magnetism

## 5.1 The Magnetic Field

Similarly to electricity, observation of magnetic phenomena goes back at least thousands of years, in many different parts of the world. There is also written evidence that ancient cultures harnessed magnetism for technology. For example, the *Sushruta Samhita* (an ancient Indian medical text written in Sanskrit) talks about using lodestones — a naturally occuring magnetic rock — to remove arrows embedded in a person's body. Nowadays magnetism is used widely in hospitals (e.g. MRI imaging), and we tend to think of the field of "medical physics" as being a new and fashionable invention. But the above example shows us that it goes back to at least ancient India! Another early use of magnetism was for navigation (i.e. compass needles, which rotate towards the north pole by following the Earth's own magnetic field). This was first documented in China, where it existed for many centuries before being brought to Europe. Other independent discoveries of magnetism occured in ancient Greece (in the town of *Magnesia*, which provides the origin of the word *magnetism*), but would also have taken place elsewhere on the globe. The reason for this is that there are certain naturally occuring minerals (e.g. magnetite), which when magnetized form the "lodestones" mentioned above. These rocks are rare, but occur in many places on the Earth's surface, hence the multiple independent discoveries of magnetism.

It is perhaps remarkable that magnetism could be harnessed for technology before a full understanding of its fundamental principles had been developed. As for electricity, this happened much later, after the notion of charged particles such as electrons had been accepted. Another key realisation was that electricity and magnetism are not separate phenomena, but have to be combined into a single description. Rather than reconstruct the tortuous history of how the subject evolved, we can take a considerable shortcut given that you know that charged particles exist! Let me then immediately tell you where magnetism comes from.

We have seen that static charges create an *electric field* $\boldsymbol{E}$ thoughout all space, such that other charges in this field feel a force. Likewise, *moving charges* generate a *magnetic field*, in addition to an electric field. We will see that magnetic fields are more complicated than the static electric fields we have seen so far, and we will worry a bit later about what the field of a moving charge actually is. For now, we will simply assume that there is such a thing as a (vector) magnetic field[1] (usually written $\boldsymbol{B}$) that can be measured, and examine the effect it has on a charged particle.

It turns out that only moving charges feel a magnetic force. This is perhaps not surprising, given that only moving charges generate magnetic fields in the first place. Experiment shows that the magnetic force on a particle of charge $q$ and velocity $\boldsymbol{v}$ in a magnetic field $\boldsymbol{B}$ is

$$\boldsymbol{F}_M = q\boldsymbol{v} \times \boldsymbol{B}. \tag{5.1}$$

We can already see that this is more complicated than the electric force. By the definition of the cross-product, the force is perpendicular both to the field, and the velocity. This is in contrast to electric forces, which are in the *same* direction as the field. Furthermore, the magnitude of the magnetic force is

$$|\boldsymbol{F}_M| = q|\boldsymbol{v}||\boldsymbol{B}|\sin\theta,$$

thus it depends on the angle between the velocity and the magnetic field. An extreme case is when a particle moves along the direction of the magnetic field $\boldsymbol{B}$, when the force experienced is zero. If electric fields are also present, the total force on a charged particle will be

$$\boldsymbol{F} = q(\boldsymbol{E} + \boldsymbol{v} \times \boldsymbol{B}). \tag{5.2}$$

This is known as the *Lorentz force*, and is a very important result. It tells us how electromagnetism influences the motion of charged particles, in full generality. By combining the Lorentz force equation with Newton's second

[1]Please note that some books and other literature refer to $\boldsymbol{B}$ as the *magnetic flux density*, reserving the term "magnetic field" for the equivalent field inside a magnetic material. Given that we will not cover magnetic materials in any detail in this book, we will not follow this convention.

law, we can completely predict how charged particles will move due to the presence of electric and magnetic fields.

Unlike the early pioneers of electricity and magnetism, you can already see from the above discussion why they have to be combined into a single theory. In particular, you may know about the theory of Special Relativity, which says that the laws of physics should be similar in all inertial frames (systems of coordinates which move at constant velocity relative to each other). However, whether or not a charge is moving or not depends on which inertial frame we are in. Thus, a charge that creates only an electric field in one frame (where it is static) will create both an electric and a magnetic field in another. In other words, electricity and magnetism have to be unified in order to be consistent with relativity! Historically, the laws of electromagnetism were discovered before those of Special Relativity. We will see why this is later on, and discuss the relationship between electromagnetism and relativity in detail in Chapter 8.

For completeness, let me point out that the SI unit of magnetic field $\boldsymbol{B}$ is the Tesla (T). Equivalently, we can write $1\ \mathrm{T} = 1\ \mathrm{NC^{-1}m^{-1}s}$.

## 5.2 Magnetic Field Lines

We saw that the concept of *field lines* was useful in talking about electric fields. These were curves in space, such that at any given point the electric field $\boldsymbol{E}$ is tangent to a field line. We can do the same for the magnetic field $\boldsymbol{B}$, although there are crucial differences compared with the electric case. Firstly, magnetic field lines *do not* represent the direction of the magnetic force. The Lorentz force equation tells us that the magnetic force is always perpendicular to the magnetic field, and thus it is perpendicular to a field line at any point.

Another difference with respect to electric field lines concerns whether field lines can begin and end. In the electric case, we saw that field lines started at positive charges, and "flowed outwards", ending either at infinity, or on negative charges. Experiment tells us that there is *no such thing* as isolated positive and negative charges for magnetism. Consider, for example, a bar magnet. This has a north and a south pole, as shown in Figure 5.1(a), and thus it looks like there are indeed two types of charge for magnetism, as there are for electricity. In the bar magnet, however, they occur together, forming a dipole. You might think that cutting the magnet in two will isolate the charges, but there is a simple experiment you can do to show that this is not the case! If you buy a bar magnet (e.g. from

Fig. 5.1 (a) A bar magnet; (b) the same bar magnet cut into two pieces.

a toy shop), and a hacksaw (probably not from a toy shop), you can easily cut the magnet into two pieces. But you will find that each piece will have a new north or south pole, as shown in Figure 5.1(b). It therefore remains true that there are no isolated magnetic charges. In other words, separating the dipole has created two new dipoles, and not isolated charges. The latter, in the case of magnetism, would be called *magnetic monopoles*, and no experiment has ever found one. However, some theories imply that such monopoles must exist somewhere in the universe, and there is also an intriguing argument first due to Dirac, that says that even if only a single magnetic monopole exists in nature, this would explain why electric charge is quantised (comes in discrete amounts). For the purposes of this book, we can simply assume that magnetic monopoles do not exist, which will be reflected accordingly in the equations we formulate for electromagnetism. It is known how to modify these equations if magnetic monopoles turn out to exist, and indeed there are current experiments around the world that are looking for them.

In the absence of magnetic monopoles, there is nothing for magnetic field lines to end on. Thus, they must form *closed loops.* An example is provided by the field due to a bar magnet, which is shown in Figure 5.2. One can demonstrate this field experimentally by, for example, scattering iron filings around the magnet. The iron is magnetised, in such a way that shards of iron line up with the magnetic field lines. This does not tell us the direction of the field, however, which loops around from the north pole of the magnet to the south pole.

As described above, magnetic field lines are very different to their electric counterparts in a number of ways. However, a common property is that if field lines are getting closer together, the field is getting stronger, and vice versa. We can see an example of this in Figure 5.2, where the field is stronger near the bar magnet, but falls off to zero at large distances.

## 5.3 Magnetic Flux

In Section 3.6, we defined the *electric flux*, as being the surface integral of the electric field through a given surface (Eq. (3.16)). This measures how

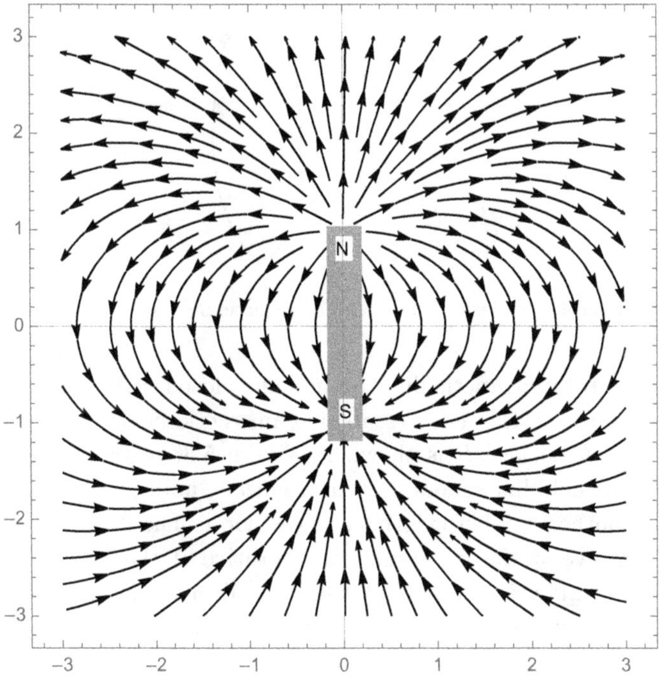


Fig. 5.2 The magnetic field due to a bar magnet.

much electric field is crossing a given surface. For closed surfaces, we saw that the total flux is related to the total enclosed charge by Gauss' Law (Eq. (3.18)). The concept of flux clearly generalised to any vector field. In particular, we can also define the *magnetic flux*

$$\Phi_B = \int\int_S \boldsymbol{B} \cdot d\boldsymbol{S}, \tag{5.3}$$

which measures how much abstract "magnetic stuff" is crossing the surface $S$, whose differential vector area on each segment of the surface is $d\boldsymbol{S}$. We can then derive a Gauss' Law for magnetism as well as electricity, and the arguments leading to this result are exactly the same in principle. However, given that there are no magnetic monopoles, there is no such thing as magnetic charge, and thus (by analogy with the electric case) we must have

$$\oint\!\!\oint_S d\boldsymbol{S} \cdot \boldsymbol{B} = 0. \tag{5.4}$$

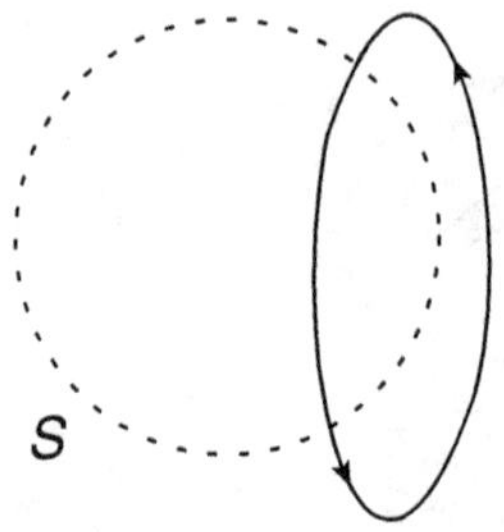

Fig. 5.3 A surface $S$ (dashed line), with a magnetic field line passing through it.

In words, the total magnetic flux through any closed surface is always zero. Another way to see this is to recall that magnetic field lines always form closed loops. A surface $S$ is shown in Figure 5.3, together with a given field line. The field line enters the surface at one point, and must leave the surface at another point if it forms a closed loop. The flux of the field will then be negative at the first point, and positive at the latter. A detailed mathematical analysis shows that this flux always cancels, for arbitrary field lines and surfaces, so that the total (*net*) flux through the surface is indeed zero.

As for Gauss' law for the electric field, Eq. (5.4) is a very general and powerful result. It is also our second Maxwell equation.

## 5.4 Motion in a Magnetic Field

In order to get a feel for how charged particles behave in magnetic fields, let us examine the simplest case: that of a particle moving in a uniform $\boldsymbol{B}$ field, and with a velocity $\boldsymbol{v}$ that is perpendicular to the field. An example is shown in Figure 5.4, where the magnetic field is pointing out of the page, and where the direction of the magnetic force $\boldsymbol{F}_M$ follows directly from Eq. (5.1). For the magnitude, the fact that the field is perpendicular to the velocity gives

$$|\boldsymbol{F}_M| = q|\boldsymbol{v}||\boldsymbol{B}| \equiv qvB,$$

where $v = |\boldsymbol{v}|$, $B = |\boldsymbol{B}|$. The force is perpendicular to the velocity, and so does no work on the particle. It follows that the speed $v$ does not change, and that the particle undergoes *uniform circular motion* with some radius $R$, as shown in Figure 5.4(b). We can find the radius by equating the expression for the magnetic force to the general expression for the centripetal force

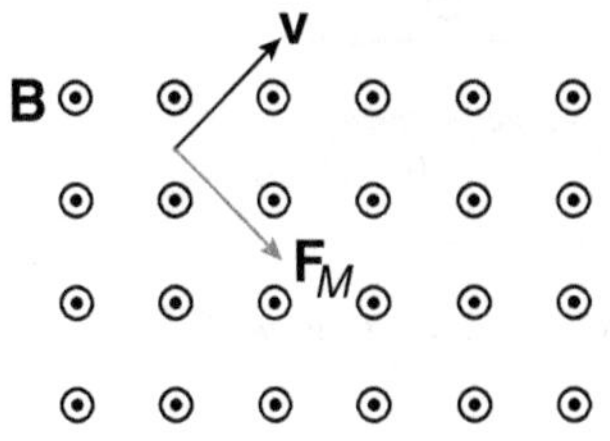


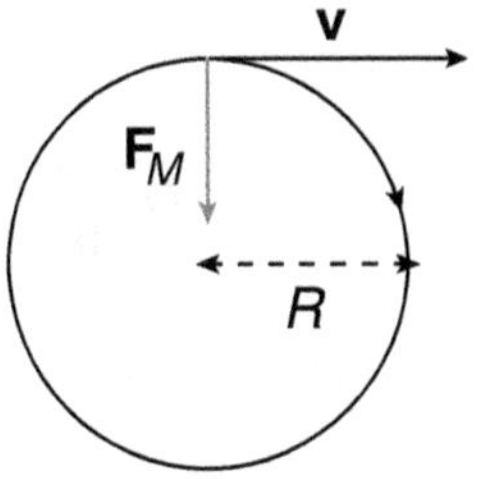


Fig. 5.4 (a) A particle moving with velocity $\boldsymbol{v}$ perpendicular to a uniform magnetic field $\boldsymbol{B}$ pointing out of the page; (b) the circular trajectory followed by such a particle.

experienced by a particle of mass $m$ undergoing uniform circular motion:

$$qvB = \frac{mv^2}{R} \quad \Rightarrow \quad R = \frac{mv}{qB}.$$

We can also find the period and frequency of the motion. For any particle undergoing uniform circular motion, the (constant) tangential speed is related to the *angular frequency* $\omega$ by

$$v = \omega R.$$

Thus, we have

$$\omega = \frac{v}{R} = \frac{qB}{m}.$$

The period is related to the angular frequency via

$$T = \frac{2\pi}{\omega} \quad \Rightarrow \quad T = \frac{2\pi m}{qB}.$$

The frequency (number of cycles per unit time) is given by

$$f = \frac{1}{T} = \frac{qB}{2\pi m}.$$

This is known as the *cyclotron* frequency, named after a certain kind of particle accelerator, that uses magnetic fields to keep charged particles moving around a ring.

We can also consider the more general case of a particle which has some component of its velocity $\boldsymbol{v}$ parallel to the uniform field $\boldsymbol{B}$. For this component, $\boldsymbol{v} \times \boldsymbol{B}$ is zero, and thus the particle moves at a constant speed parallel to the field, whilst undergoing uniform circular motion perpendicular to $\boldsymbol{B}$.

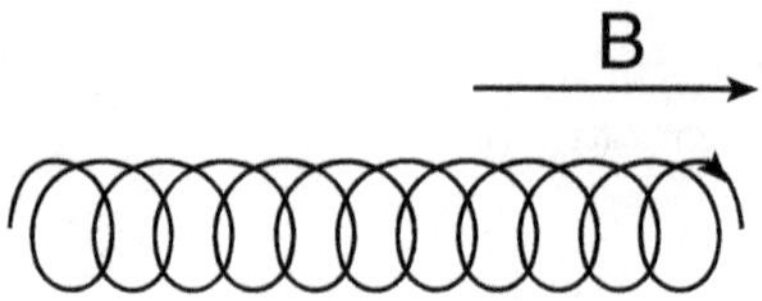


Fig. 5.5 The helical trajectory followed by a charged particle with non-zero velocity parallel to a uniform magnetic field.

The combination of these two effects is a *helical* or *spiral trajectory*, as shown in Figure 5.5.

## 5.5 The Magnetic Field of a Moving Charge

Earlier, we said that magnetic fields are generated by moving charges, but did not make this more precise. We shall now do so. Consider a particle of charge $q$ at the origin, moving with velocity $\boldsymbol{v}$. From experiment, the magnetic field at displacement $\boldsymbol{x}$ is found to satisfy

$$\boldsymbol{B} \propto q \frac{\boldsymbol{v} \times \hat{r}}{r^2},$$

where $\hat{r}$ is a unit vector in the radial direction, and $r$ the radial distance. Note that this expression behaves like $r^{-2}$, as in the case of the electric field. The direction, however, is very different! It is perpendicular to both $\boldsymbol{v}$ and $\hat{r}$, and the magnitude of the field will satisfy

$$|\boldsymbol{B}| \propto \frac{q|\boldsymbol{v}| \sin\theta}{r^2},$$

where $\theta$ is the angle between $\boldsymbol{v}$ and the direction $\hat{r}$. For historical reasons, the constant of proportionality in the above equation is usually written as

$$\frac{\mu_0}{4\pi},$$

where $\mu_0$ is called the *permeability of free space*. Also, we may write the unit vector in the radial direction at point $\boldsymbol{x}$ as

$$\hat{\boldsymbol{r}} = \frac{\boldsymbol{x}}{|\boldsymbol{x}|}, \quad r = |\boldsymbol{x}|,$$

so that one finally obtains

$$\boldsymbol{B} = \frac{\mu_0 q}{4\pi} \frac{\boldsymbol{v} \times \boldsymbol{x}}{|\boldsymbol{x}|^3}. \tag{5.5}$$

In general, we might have a particle at position $\boldsymbol{x}'$ rather than the origin. Equation (5.5) then implies that the field at point $\boldsymbol{x}$ due to the moving charge at $\boldsymbol{x}'$ is given by

$$\boldsymbol{B} = \frac{\mu_0 q}{4\pi} \frac{\boldsymbol{v} \times (\boldsymbol{x} - \boldsymbol{x}')}{|\boldsymbol{x} - \boldsymbol{x}'|^3}. \tag{5.6}$$

This result is for a single point charge, and we could have more than one charge in general. For electric fields, we saw earlier that we could find the field due to multiple charges by superposing the electric fields due to each individual charge. This followed from the fact that the electric force was *linear* in the electric field, so that the principle of superposition of forces to get the net force on a particle translated directly into superposition of the fields. Similar arguments apply to the case of the magnetic force, given that this is linear in the magnetic field $\boldsymbol{B}$. Thus, the magnetic field due to multiple charges $q_i$ obeys the superposition principle

$$\boldsymbol{B} = \sum_i \boldsymbol{B}_i, \tag{5.7}$$

where $\boldsymbol{B}_i$ is the field due to the single charge $q_i$.

## 5.6 The Biot–Savart Law

In applications involving moving charges, it is more common to talk about the *current* in, e.g. a wire, rather than the velocity of moving charges directly. In such cases, we can rewrite Eq. (5.5) to give the field due to a small segment $\boldsymbol{dl}$ of a current-carrying wire at the origin. Such a segment will have a volume $Adl$, where $dl = |\boldsymbol{dl}|$ and $A$ is the cross-sectional area of the wire. The total amount of charge in this segment is then given by

$$dQ = nqAdl,$$

where $n$ is the number of mobile charge carriers (each of charge $q$) per unit volume. The charges themselves move with drift velocity $\boldsymbol{v}_d$, thus generate a total magnetic field at $\boldsymbol{x}$

$$d\boldsymbol{B} = \frac{\mu_0 dQ}{4\pi} \frac{\boldsymbol{v}_d \times \boldsymbol{x}}{|\boldsymbol{x}|^3}.$$

In this formula, we may write

$$dQ\boldsymbol{v}_d = dQ|\boldsymbol{v}_d|\hat{\boldsymbol{t}},$$

where $\hat{\boldsymbol{t}}$ is a unit vector tangent to the curve of the wire. We can also write

$$d\boldsymbol{l} = dl\hat{\boldsymbol{t}}$$

(i.e. the vector segment $d\boldsymbol{l}$ is just the infinitesimal length of the segment, multiplied by the unit tangent vector). Thus, we have

$$dQ\boldsymbol{v}_d = nqA|\boldsymbol{v}_d|d\boldsymbol{l} = Id\boldsymbol{l}, \tag{5.8}$$

where we have used Eq. (4.6) to recognise the combination

$$nqA|\boldsymbol{v}_d|$$

as the current through the wire. Putting everything together, we find that the magnetic field at point $\boldsymbol{x}$ due to a segment of wire $d\boldsymbol{l}$ at the origin carrying current $I$ (in the direction of $d\boldsymbol{l}$) is

$$d\boldsymbol{B} = \frac{\mu_0 I}{4\pi}\frac{d\boldsymbol{l}\times\boldsymbol{x}}{|\boldsymbol{x}|^3}. \tag{5.9}$$

This is known as the *Biot–Savart law*, and proves very useful when analysing the magnetic field due to various combinations of current-carrying wires. It is straightforward to generalise Eq. (5.9) to the case when the segment of wire is at a point $\boldsymbol{x}'$ rather than the origin. Then the field at point $\boldsymbol{x}$ due to this segment is

$$d\boldsymbol{B} = \frac{\mu_0 I}{4\pi}\frac{d\boldsymbol{l}\times(\boldsymbol{x}-\boldsymbol{x}')}{|\boldsymbol{x}-\boldsymbol{x}'|^3}. \tag{5.10}$$

As an example of how to use this equation, consider the field due to an infinite wire, carrying current $I$, as shown in Figure 5.6. One may consider the field at a point $P$ a perpendicular distance $R$ from the wire, as shown in the figure, and the Biot–Savart law gives the result as

$$d\boldsymbol{B} = \frac{\mu_0 I}{4\pi}\frac{d\boldsymbol{l}\times\hat{\boldsymbol{r}}}{r^2},$$

where

$$\hat{\boldsymbol{r}} = \begin{pmatrix}\sin\theta\\ \cos\theta\\ 0\end{pmatrix}$$

is a unit vector pointing from the segment $d\boldsymbol{l}$ towards the point $P$, and

$$d\boldsymbol{l} = \begin{pmatrix}0\\ dl\\ 0\end{pmatrix},$$

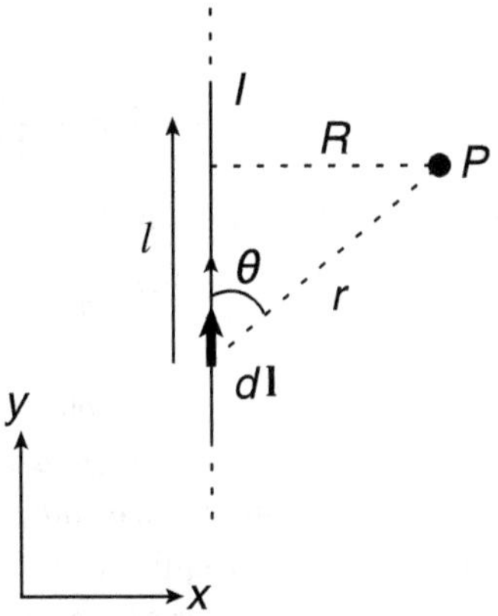


Fig. 5.6 A point $P$ a perpendicular distance $R$ from an infinite wire carrying current $I$.

given that we have placed the wire along the $y$-direction. One then has

$$d\boldsymbol{l} \times \hat{\boldsymbol{r}} = \begin{pmatrix} 0 \\ dl \\ 0 \end{pmatrix} \times \begin{pmatrix} \sin\theta \\ \cos\theta \\ 0 \end{pmatrix} = \begin{pmatrix} 0 \\ 0 \\ -dl\sin\theta \end{pmatrix},$$

which points into the page (i.e. for a right-handed coordinate system, the $z$-direction must be pointing out of the page, so that into the page is the negative $z$-direction). The length $r$ varies as the segment $d\boldsymbol{l}$ moves along the wire, as does the angle $\theta$. These two variables are related by

$$R = r\sin\theta \quad \Rightarrow \quad r = \frac{R}{\sin\theta}.$$

We can then write the magnitude of the field above as

$$\begin{aligned} |d\boldsymbol{B}| &= \frac{\mu_0 I}{4\pi}\left(\frac{R}{\sin\theta}\right)^{-2} dl\sin\theta \\ &= \frac{\mu_0 I dl}{4\pi R^2}\sin^3\theta. \end{aligned}$$

Also, $dl$ is the change in $l$, where the latter is a coordinate measuring the length along the wire from some arbitrary starting point, as shown in Figure 5.6. It follows that $l$ and the angle $\theta$ are not independent of each other: both of them vary as we move the segment $d\boldsymbol{l}$ along the wire, and the figure gives

$$l = \frac{R}{\tan\theta} \quad \Rightarrow \quad dl = \frac{dl}{d\theta}d\theta = -\frac{R}{\sin^2\theta}d\theta,$$

where we have used the chain rule. Thus, one has

$$|d\boldsymbol{B}| = -\frac{\mu_0 I \sin\theta}{4\pi R}d\theta,$$

To explain the minus sign, note that $l$ is increasing as we move from the bottom of the wire to the top. However, $\theta$ is *decreasing*: it is $\pi$ at the bottom of the wire, and 0 at the top. Hence, $d\theta$ is negative in the above formula, such that $|d\boldsymbol{B}|$ is positive as required. We can then find the total magnitude of the field by integrating over all values of $\theta$:

$$\begin{aligned} B &= -\int_{\pi}^{0} d\theta \frac{\mu_0 I}{4\pi}\frac{\sin\theta}{R} \\ &= \frac{\mu_0 I}{4\pi R}\int_0^{\pi} d\theta \sin\theta \\ &= \frac{\mu_0 I}{2\pi R}. \end{aligned} \tag{5.11}$$

Another way to think about the direction of the field is to consider looking down on the wire, as shown in Figure 5.7. The field lines (curves to which $\boldsymbol{B}$ is tangent at any point in space) are circles, centred on the wire. It is hopefully straightforward to check that the field direction is consistent with the direction we found above. Furthermore, we can note that there is a shortcut to obtaining this direction, namely using a *right-hand rule*: upon placing the thumb in the direction of the current, the fingers of the right-hand curl in the direction of the magnetic field.

## 5.7 Force on a Wire

The magnetic force on a particle of charge $q$ is given by Eq. (5.1). It then follows from Eq. (5.8) that, if a segment of wire $\boldsymbol{dl}$ carrying current $I$ is placed in a magnetic field, it feels a force due to the field $\boldsymbol{B}$ at that point given by

$$d\boldsymbol{F} = dQ\boldsymbol{v}_d \times \boldsymbol{B} = I\boldsymbol{dl} \times \boldsymbol{B}.$$

Now consider two parallel wires carrying currents $I_1$ and $I_2$, as shown in Figure 5.8. One may further consider segments of each wire, $\boldsymbol{dl}_1$ and $\boldsymbol{dl}_2$, separated by a perpendicular distance $R$. From Eq. (5.11), the segment $d\boldsymbol{l}_2$ of wire 2 experiences a magnetic field (due to wire 1) with magnitude

$$|\boldsymbol{B}_1| = \frac{\mu_0 I_1}{2\pi R},$$

where the direction is into the page. Likewise, segment $d\boldsymbol{l}_1$ of wire 1 experiences a magnetic field $\boldsymbol{B}_2$ (due to wire 2) with magnitude

$$|\boldsymbol{B}_2| = \frac{\mu_0 I_2}{2\pi R},$$

and directed *into* the page. Thus, each segment of wire 2 feels a force $d\boldsymbol{F}_2$ towards wire 1, with magnitude

$$|d\boldsymbol{F}_2| = (I_2 d\boldsymbol{l}_2)\frac{\mu_0 I_1}{2\pi R} = \frac{\mu_0 I_1 I_2}{2\pi R} dl_2.$$

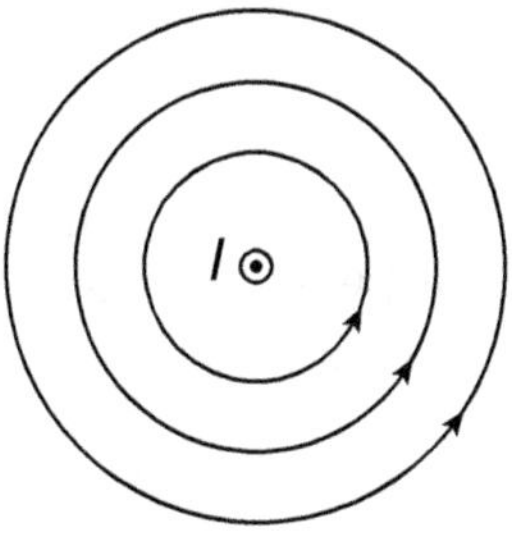


Fig. 5.7 The magnetic field lines around a current-carrying wire, where the current is flowing out of the page.

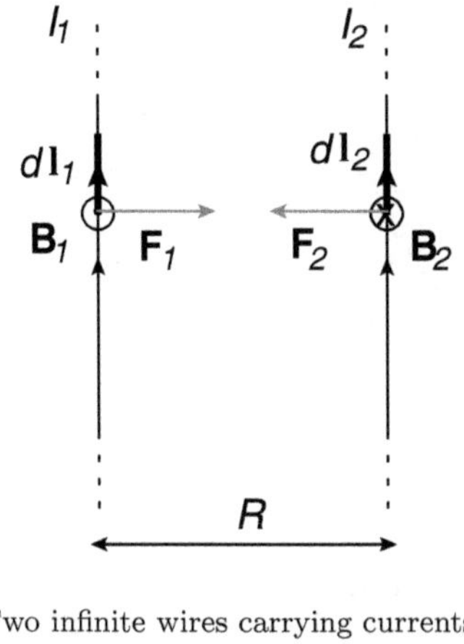


Fig. 5.8 Two infinite wires carrying currents $I_1$ and $I_2$.

Similarly, each segment of wire 1 feels a force towards wire 2, with magnitude

$$|d\boldsymbol{F}_1| = (I_1 dl_1)\frac{\mu_0 I_2}{2\pi R} = \frac{\mu_0 I_1 I_2}{2\pi R} dl_1.$$

Each wire then feels a force per unit length

$$\frac{dF}{dl} = \frac{\mu_0 I_1 I_2}{2\pi R}. \tag{5.12}$$

Here we have taken the currents to be in the same direction. If the wires carry opposing currents, the force would be repulsive rather than attractive. Furthermore, this is the total electromagnetic force between the wires: there is no electric force, assuming each wire carries no *net* charge. The magnetic force depends only on the charge that is moving, and it is only one type of charge that moves in the material.

## 5.8 Torque on a Current Loop

Consider a rectangular loop of current in a uniform $\boldsymbol{B}$ field perpendicular to the loop, as shown in Figure 5.9(a), and where $a$ and $b$ denote the lengths of the sides. From the above result

$$d\boldsymbol{F} = I d\boldsymbol{l} \times \boldsymbol{B},$$

it follows that each side of the loop feels a force directed outwards from the centre of the loop as shown, and where

$$|\boldsymbol{F}_A| = |\boldsymbol{F}_C| = IBa, \quad |\boldsymbol{F}_B| = |\boldsymbol{F}_D| = IBb,$$

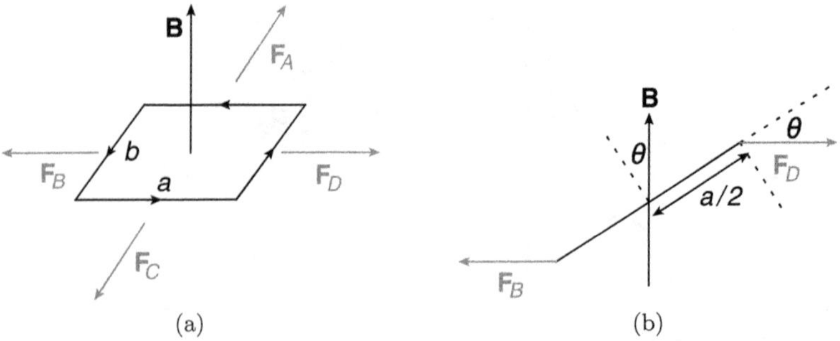


Fig. 5.9 (a) A current loop in a perpendicular uniform magnetic field; (b) a current loop whose normal is at an angle $\theta$ to a uniform magnetic field (seen sideways on).

with $B \equiv |\boldsymbol{B}|$, and $I$ the current. Due to the fact that the forces are equal and opposite on opposite sides, the net force on the loop is zero, and there is merely a tendency for the loop to expand (which would be counteracted by tension forces in the case of a real loop, assuming it remained intact).

Now consider a similar loop which is placed at an angle to the field, such that the angle between the normal of the loop and the magnetic field is $\theta$. We depict this in Figure 5.9(b), where the loop is viewed sideways on. Now there is still no net force on the loop. However, the forces $\boldsymbol{F}_B$ and $\boldsymbol{F}_D$ create a torque about the centre of the loop. The component of $\boldsymbol{F}_D$ perpendicular to the sides of the loop (i.e. the component in Figure 5.9 that contributes to the torque) is given by

$$|\boldsymbol{F}_D| \sin\theta.$$

Furthermore, $\boldsymbol{F}_B$ and $\boldsymbol{F}_D$ are equal and opposite, and so both give the same contribution to the total torque, whose magnitude is thus

$$\begin{aligned}|\boldsymbol{G}| &= 2 \times \frac{a}{2} \times |\boldsymbol{F}_D| \sin\theta \\ &= IBab \sin\theta \\ &= IBA \sin\theta,\end{aligned}$$

where $A$ is the area of the loop. It is conventional to define the *magnetic dipole moment vector*

$$\mu_B = I\boldsymbol{S}, \tag{5.13}$$

where $\boldsymbol{S}$ is the vector area of the loop i.e. a vector whose magnitude is the area of the loop, and whose direction is given by a right-hand rule: if the thumb of the right-hand points perpendicularly outwards from the loop (in the direction of $\boldsymbol{S}$), the fingers should curl in the direction of the current around the loop. For example, in Figure 5.9(a), the vector area points in the direction of $\boldsymbol{B}$. Given Eq. (5.13), the torque on the current loop can be written as

$$\boldsymbol{G} = \mu_B \times \boldsymbol{B}. \tag{5.14}$$

It is straightforward to verify that this produces the magnitude we obtained above. Furthermore, in our example, the torque vector points into the page, leading to a clockwise rotation as expected.

If there are $N$ current loops on top of each other, the force on each edge is simply $N$ times larger, and thus the torque is too. But you may be wondering why I am telling you all this. Firstly, we saw that there are things called dipoles in electricity, so it is useful to know that there is some sort of analogue in magnetism. Secondly, there are a large number of practical applications of magnetic dipoles: they are used, e.g. in loudspeakers, and in galvanometers for measuring current.

Thirdly, telling you about magnetic dipoles allows me to be able to explain what exactly is going on inside a bar magnet, which you may have been curious about (if not, feel free to ignore the following paragraph). We saw earlier that when you cut a bar magnet, you always get a new north or south pole, so that the smaller piece you are left with is still a dipole. But what if you kept cutting the magnet, so that the remaining piece got smaller and smaller? Surely, at some point, you would be left with a single magnetic charge (monopole)? The answer is no, due to the fact that the magnetism arises from the electrons themselves that are in the bar magnet! Electrons are so-called *fundamental particles*: they are pointlike, and we cannot chop them into smaller pieces. They have a weird quantum property called *spin*, that means that they carry angular momentum. To see why this is weird, note that a classical object (e.g. a book) can only have angular momentum about its centre of mass if it has a nonzero size: we need bits of the object that are displaced from the centre of the mass to get any angular momentum, by definition. However, electrons are *pointlike*, but still look like they have angular momentum. A charged particle that is apparently "spinning" looks like a small loop of current. Thus, it is a magnetic dipole! What happens in the simplest magnetic materials, when they are "magnetised", is that these dipoles all line up throughout the solid, creating a net magnetic dipole which you see as the north and south poles of a bar magnet. The fact that you cannot separate the charges amounts to the fact that the fundamental objects causing the magnetism (electrons) are themselves magnetic dipoles, and not monopoles.

## 5.9 Ampère's Law

In the previous sections, we have seen how we can find the magnetic field generated by current distributions using the Biot–Savart law. Whilst this is in principle always possible for arbitrary current distributions, it can be very cumbersome in practice. This is analogous to how, in electricity, we could in principle find the electric field using Coulomb's law. However, in many situations it was much simpler to use Gauss' law, which ended up being a more fundamental equation for the electric field. In this section, we apply a similar idea to magnetism.

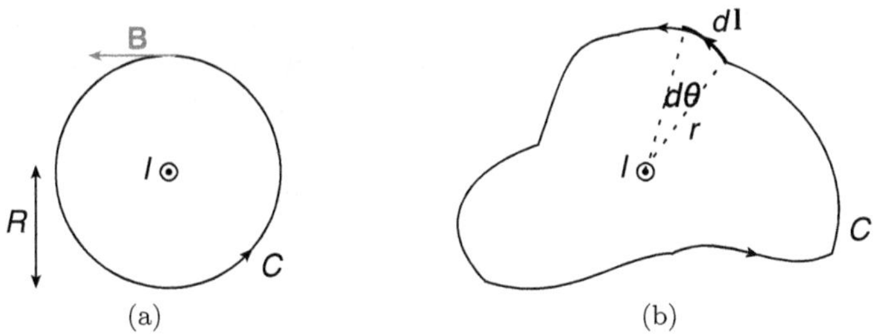


Fig. 5.10 (a) A circular contour around a current carrying wire; (b) an arbitrary contour looks locally like a circular one.

Consider a circular loop $C$ around a wire carrying current $I$, as shown in Figure 5.10(a). As we have seen above, the magnetic field on each segment of the line is tangential to the circle, and thus on an arbitrary segment $\boldsymbol{dl}$, one has

$$\boldsymbol{B} \cdot \boldsymbol{dl} = |\boldsymbol{B}| dl = \frac{\mu_0 I}{2\pi R} dl.$$

The total line integral around the closed curve is then

$$\begin{aligned}\oint_C \boldsymbol{B} \cdot \boldsymbol{dl} &= \int dl \frac{\mu_0 I}{2\pi R} \\ &= \frac{\mu_0 I}{2\pi R} \int dl,\end{aligned}$$

where we have used Eq. (5.11), and also the fact that the radius $R$ is constant, so can be taken outside the integral. On the left-hand side of the above equation, we have appended a loop to the integral symbol, which is the conventional notation for the line integral around a closed loop. Some people omit the symbol, but it is sometimes useful as it reminds us which equations work only for closed loops, and which are more general.

The integral over the length above just gives the total length of the circular contour, i.e.

$$\int dl = 2\pi R,$$

and we therefore find

$$\oint \boldsymbol{B} \cdot \boldsymbol{dl} = \frac{\mu_0 I}{2\pi R} 2\pi R = \mu_0 I.$$

In fact, this result turns out to be much more general than it first appears. Consider an arbitrary (non-circular) closed contour, an example of which is shown in Figure 5.10(b). A given segment $\boldsymbol{dl}$ is shown, and provided this is small enough, it will look locally like a small arc of a circle[2] of radius $r$, subtending a small angle $d\theta$. On this segment, we can use our previous result for circular contours to say that the field will be parallel to $\boldsymbol{dl}$, with

$$|\boldsymbol{B}| = \frac{\mu_0 I}{2\pi r}.$$

Also, the length of the segment is (from the geometry of the figure)

$$dl \equiv |\boldsymbol{dl}| = r d\theta.$$

Thus, we have

$$\boldsymbol{B}\cdot\boldsymbol{dl} = |\boldsymbol{B}||\boldsymbol{dl}| = \frac{\mu_0 I}{2\pi r} r d\theta = \frac{\mu_0 I}{2\pi} d\theta.$$

Note that the radial distance $r$ has cancelled in this formula, and thus we would have obtained the same result, for *any* segment on the contour! We can then sum over all such segments by integrating over all values of the angle $\theta$. Taking, e.g. the horizontal direction to the right as defining $\theta = 0$ in polar coordinates, this will vary from 0 to $2\pi$ (a complete circle). We thus get

$$\begin{aligned}\oint_C \boldsymbol{B}\cdot\boldsymbol{dl} &= \int_0^{2\pi} d\theta \frac{\mu_0 I}{2\pi} \\ &= \mu_0 I,\end{aligned}$$

exactly the same result as before. As well as choosing arbitrary contours $C$, we can also consider more than one wire (or other source of current) inside the loop. Then we can use the principle of superposition on each segment $\boldsymbol{dl}$, which says that the total magnetic field at that point is just the sum of the individual fields due to each separate current $I_i$:

$$\boldsymbol{B} = \sum_i \boldsymbol{B}_i = \sum_i \frac{\mu_0 I_i}{2\pi r} = \frac{\mu_0}{2\pi r}\sum_i I_i.$$

[2]If we zoom in far enough, any curved line or circle will look like a small piece of a straight line, so that it does not matter which way a given segment is curving.

The above argument then goes through unchanged, provided we replace $I$ with

$$I_{\text{tot.}} = \sum_i I_i.$$

Finally, we arrive at the general result

$$\oint_C \boldsymbol{B} \cdot d\boldsymbol{l} = \mu_0 I_{\text{tot.}}. \tag{5.15}$$

In words: the line integral of the magnetic field $\boldsymbol{B}$ around an arbitrary closed contour is equal to the permeability of free space, multiplied by the total current enclosed by the loop. This result is known as *Ampère's law*, and is very general and powerful. It works for completely general current distributions (including continuous distributions), and allows us to more straightforwardly calculate magnetic fields, even in quite complicated situations. Just as Gauss' law for electric fields allowed us to rederive Coulomb's law, we can take Ampère's law to be the fundamental equation describing how $\boldsymbol{B}$ fields are generated by sources. Indeed, it forms one of the Maxwell equations, albeit in a modified form to be discussed later.

Note that Ampère's law tells us that magnetic fields are very different in nature to electric fields. In electrostatics, the electric field satisfied

$$\oint_C \boldsymbol{E} \cdot d\boldsymbol{l} = 0,$$

due to the fact that $\boldsymbol{E}$ is a conservative field (i.e. the total change in potential around any closed loop is zero). By contrast, Ampère's law tells us that the magnetic field $\boldsymbol{B}$ is *not conservative*, and that there is no well-defined scalar potential for magnetism. We return to this point in Chapter 8.

To clarify how to use Ampère's law, let us consider a couple of examples.

### 1. A conducting cylinder

Consider an infinite conducting cylinder of radius $R$, with uniform current density $J$. We can then ask: what is the magnetic field a perpendicular distance $r$ from the axis? In principle, we could divide such a cylinder into lots of tiny current-carrying segments and use the Biot–Savart law. However, this would clearly be horrendously complicated, as the case of a single wire above was bad enough! The analysis is instead much simpler if we use Ampère's law. We consider a section of the cylinder in Figure 5.11. By symmetry (and from the right-hand rule we found for the current in a wire), the field will point in the azimuthal ($\phi$) direction, indicated in the

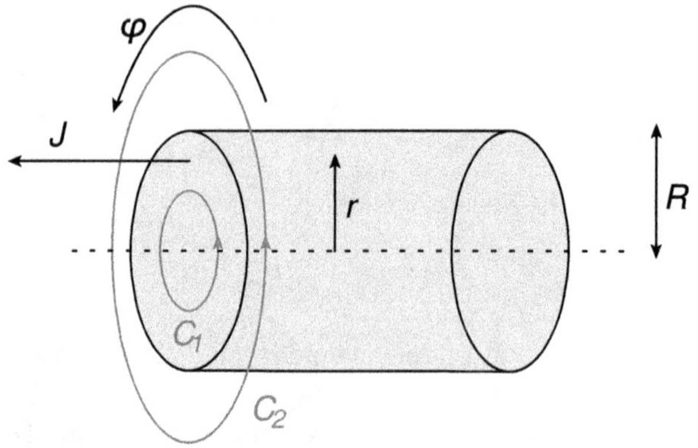


Fig. 5.11 A conducting cylinder, with uniform current density $J$.

figure, and will depend only on the perpendicular distance $r$. For Ampère's law, it is easiest to use a contour that reflects the cylindrical symmetry of the situation, namely a circle at fixed $r$. On such a contour, $\boldsymbol{B}$ is parallel to $\boldsymbol{dl}$ everywhere, and $|\boldsymbol{B}|$ is constant.

Inside the cylinder, we can use a contour $C_1$ as shown in Figure 5.11. This gives

$$\oint_{C_1} \boldsymbol{B} \cdot \boldsymbol{dl} = \oint_{C_1} dlB = B \oint dl = 2\pi r B,$$

where we have used the fact that $\boldsymbol{B} \parallel \boldsymbol{dl}$, with constant magnitude. The total current enclosed by the contour $C_1$ is

$$\pi r^2 J,$$

(i.e. the current density multiplied by the area enclosed by $C_1$). Ampère's law then gives

$$2\pi r B = \mu_0 \pi r^2 J \quad \Rightarrow B = \frac{\mu_0}{2} J r.$$

Outside the cylinder, we can use the contour $C_2$ shown in Figure 5.11, which similarly to above gives

$$\oint_{C_2} \boldsymbol{B} \cdot \boldsymbol{dl} = 2\pi r B.$$

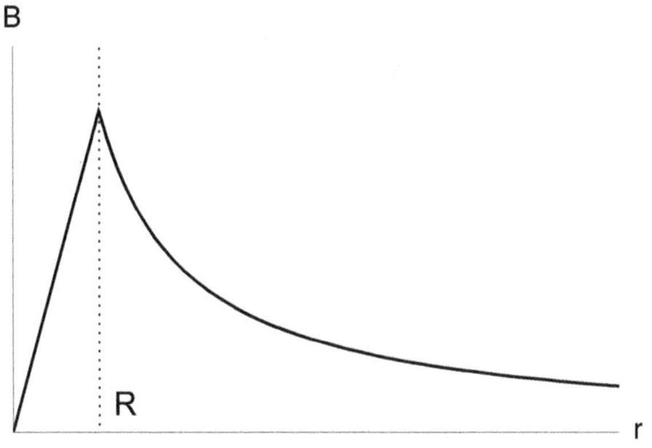


Fig. 5.12 The magnitude of the magnetic field due to a uniform conducting cylinder of radius $R$, as a function of perpendicular distance $r$ from its axis.

Now, however, Ampère's law gives

$$2\pi r B = \mu_0 I,$$

where $I$ is the total current carried by the wire. We thus have

$$B = \begin{cases} \frac{\mu_0 J r}{2}, & r < R \\ \frac{\mu_0 I}{2\pi r}, & r > R. \end{cases}$$

A sketch of the result is shown in Figure 5.12. Interestingly, the result for $r > R$ agrees with our previous result derived (much more cumbersomely) from the Biot–Savart law. Thus, outside the cylinder, it looks as if all the current were concentrated in a thin wire along its axis. This is reminiscent of how, in electrostatics, a charged sphere looks effectively like a point charge if we are outside the sphere.

### 2. A solenoid

A *solenoid* consists of a series of current-carrying coils of wire. For a single coil, $\boldsymbol{B}$ inside the loop is perpendicular to the coil (e.g. Figure 5.9(a)). Thus, the field inside a solenoid is parallel to the axis, as shown in Figure 5.13, which shows a section of the device. To find the field, we can then use Ampère's law applied to the contour, and composed of four parts:

$$\oint_C \boldsymbol{B} \cdot d\boldsymbol{l} = \int_{C_1} \boldsymbol{B} \cdot d\boldsymbol{l} + \int_{C_2} \boldsymbol{B} \cdot d\boldsymbol{l} + \int_{C_3} \boldsymbol{B} \cdot d\boldsymbol{l} + \int_{C_4} \boldsymbol{B} \cdot d\boldsymbol{l} = \mu_0 I_{\text{tot.}}.$$

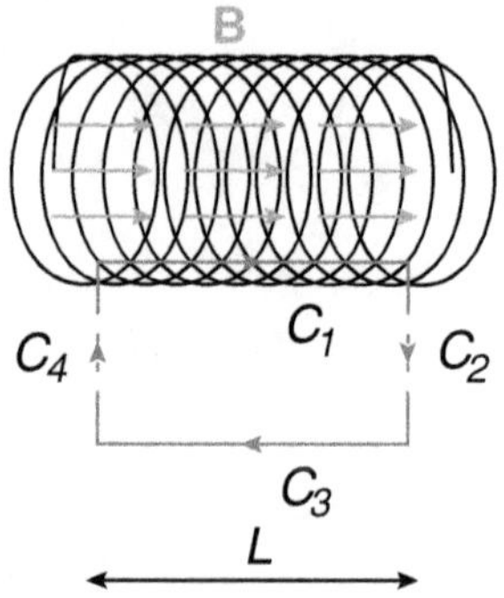


Fig. 5.13 A solenoid, consisting of a number of coils of wire.

It is convenient to take the contours $C_2$ and $C_4$ to be very long, so that $C_3$ is pushed to infinity (this is indicated by the dotted lines in the figure). The magnetic field should fall away to zero here, so that we then have

$$\int_{C_3} \boldsymbol{B} \cdot d\boldsymbol{l} = 0.$$

Also, $\boldsymbol{B}$ is perpendicular to the segments $C_2$ and $C_4$, which in turn implies

$$\int_{C_2} \boldsymbol{B} \cdot d\boldsymbol{l} = \int_{C_4} \boldsymbol{B} \cdot d\boldsymbol{l} = 0.$$

Finally, there is the contribution from the contour $C_1$. For a long solenoid, the field will be uniform along the contour, so that we get

$$\int_{C_1} \boldsymbol{B} \cdot d\boldsymbol{l} = \int_{C_1} |\boldsymbol{B}| dl = |\boldsymbol{B}| \int dl = |\boldsymbol{B}| L,$$

which (given the other portions of the total contour $C$ all give zero) corresponds to the total line integral of $\boldsymbol{B}$ around the closed loop. Ampère's law relates this to the total current enclosed by the loop, which will be given by $NI$, where $I$ is the current carried by the wire forming the solenoid, and $N$ the number of coils in length $L$. We therefore find

$$|\boldsymbol{B}| = \frac{\mu_0 NI}{L}, \tag{5.16}$$

or

$$|\boldsymbol{B}| = \mu_0 nI, \tag{5.17}$$

where $n$ is the number of turns per unit length. Note that this is indeed uniform throughout the solenoid: it does not depend on where we put $C_1$, so long as it is inside the coils.

## 5.10 Electromagnetic Induction

Ampère's law is one of the key equations defining how magnetic fields are generated by currents. There is another important phenomenon involving magnetic fields, that we will explore in this section. First, let me ask you to take something on trust if you have not seen it before: experiment shows that moving a bar magnet through a coil of wire produces an electric current $I$ in the coil. This effect is called *electromagnetic induction*, and if the coil has some resistance, then there will be an *induced EMF* with magnitude

$$|\mathcal{E}| = IR.$$

In terms of the fields involved, the fact that magnets induce a current suggests that the magnetic flux through the coil is important. However, a stationary magnet is not found to induce a current. Thus, $I$ must depend only on *changes* in the magnetic flux through the coil. Detailed experiments can be done that indeed establish the following relationship between the induced EMF $\mathcal{E}$ and the magnetic flux $\Phi_B$ through a loop of wire:

$$\mathcal{E} = -\frac{d\Phi_B}{dt}. \tag{5.18}$$

That is, the induced EMF in a loop of wire is directly related to the *rate of change* of magnetic flux through the loop. This result is known as *Faraday's law*, and tells us that there is indeed no induced EMF if $\Phi_B$ is not changing, consistent with observations. In order to use Eq. (5.18), we must define what is meant by positive EMF $\mathcal{E}$ in the loop. This is defined using a right-hand rule: if the thumb points parallel to the vector area of a loop, the fingers curl in the direction of a *positive* induced current $I$. An example is shown in Figure 5.14, which considers a coil with vector area $\boldsymbol{S}$, such that the rate of change of flux is assumed to be positive, By Eq. (5.18), this induces a *negative EMF*, leading to a current flowing in the clockwise direction. By the Biot–Savart law, a given segment $\boldsymbol{dl}$ of the loop will create an *induced magnetic field*

$$\boldsymbol{B}_i \sim I\boldsymbol{dl} \times \boldsymbol{r}$$

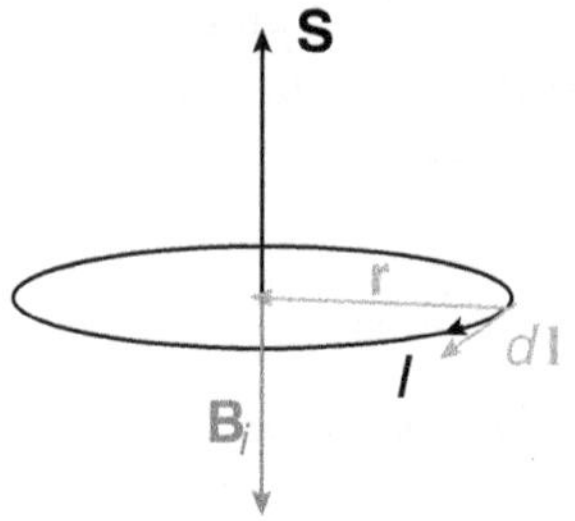


Fig. 5.14 A positive change of flux through the loop induces a current $I$ in the direction shown. This in turn induces a magnetic field that opposes the change in the flux.

through the loop, where $\boldsymbol{r}$ is a vector from the segment to a point inside the loop. The direction of this induced field is opposite to $\boldsymbol{S}$, and thus *opposes* the increasing flux. We can carry out a similar analysis for the case where the rate of change of flux is negative. Then the induced field is parallel to $\boldsymbol{S}$, which acts to *oppose* the *decreasing* flux. This is the physical origin of the minus sign in Faraday's law, and by itself is known as *Lenz's law*: induced currents/EMFs always act to oppose the change that produced them. Were the opposite true (i.e. if the induced magnetic field added to the magnetic flux through a loop), we could create arbitrarily strong magnetic fields, carrying more and more energy. This clearly violates the conservation of energy. The presence of an EMF implies that there must be an *induced electric field* $\boldsymbol{E}$, such that

$$\mathcal{E} = \oint_C \boldsymbol{E} \cdot d\boldsymbol{l}, \tag{5.19}$$

where $C$ denotes a contour around the loop. This electric field is *not* conservative, as in the case of fields from static charges. This is not surprising: the induced electric field is caused by a changing magnetic flux, and we know that $\boldsymbol{B}$ is not conservative either. Equation (5.19) allows us to rewrite Faraday's law in terms of the electric field directly:

$$\oint \boldsymbol{E} \cdot d\boldsymbol{l} = -\frac{d\Phi_B}{dt}. \tag{5.20}$$

In words: the line integral of the electric field around a closed loop $C$ is minus the rate of change of magnetic flux through the loop. In this equation, we can choose *any* surface spanned by $C$ to calculate the flux. Imagine that we have two different surfaces, as shown in Figure 5.15. Provided there are no sources of a given vector field in between $S_1$ and $S_2$, all of the flux

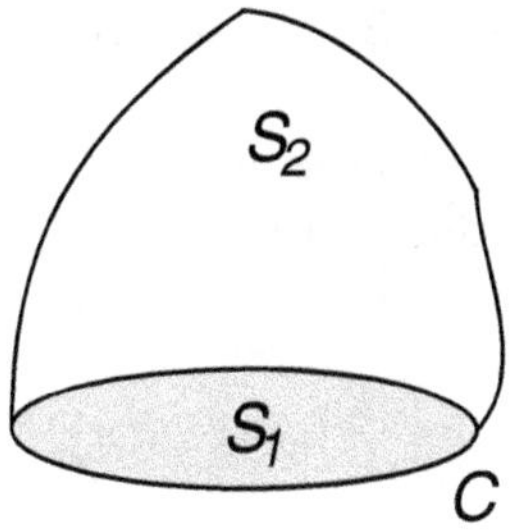


Fig. 5.15 Two surfaces spanning the same curve $C$.

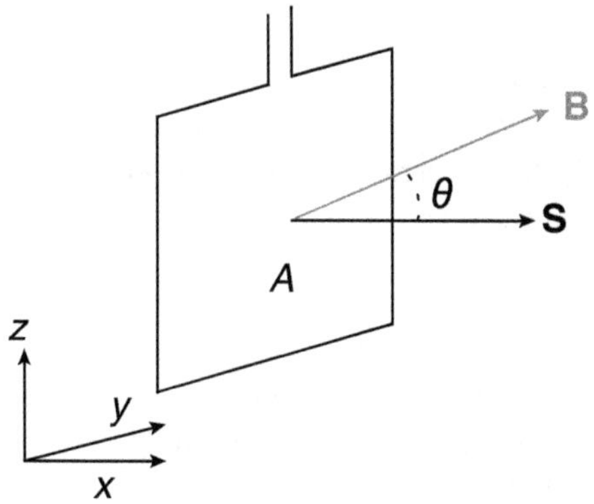


Fig. 5.16 A loop of wire enclosing area $A$, in a uniform magnetic field $\boldsymbol{B}$.

entering the surface $S_1$ must leave through the surface $S_2$. For the magnetic field, there are no sources (magnetic monopoles), and thus the fluxes through the two surfaces must indeed be equal.

Faraday's law in the form of Eq. (5.20) is taken to be a fundamental principle of electromagnetism. It plays a key role in Maxwell's equations, as we will see. There are also many practical applications, and an example is a device called an *alternator*. Consider a uniform magnetic field passing through a loop of wire in the $(y, z)$ plane, as shown in Figure 5.16, and let us take the field to be given by

$$\boldsymbol{B} = B_0 \begin{pmatrix} \cos\theta \\ \sin\theta \\ 0 \end{pmatrix}.$$

Then the magnetic flux through the loop is given by

$$\Phi_B = \boldsymbol{B} \cdot \boldsymbol{S} = B_0 A \cos\theta,$$

where $A$ is the area enclosed. Now assume that whatever is causing the field (e.g. a bar magnet) is rotating so that $\theta = \omega t$, for some constant angular velocity $\omega$. From the above expression, the flux will be given by

$$\Phi_B = B_0 A \cos(\omega t),$$

and Faraday's law then implies that an EMF is induced in the loop:

$$\mathcal{E} = -\frac{d\Phi_B}{dt} = -B_0 A \frac{d}{dt}\cos(\omega t) = B_0 A \omega \sin(\omega t).$$

The device thus acts as a source of sinusoidally alternating voltage, or equivalently generates *alternating current* (AC). Such a time-varying current is widely used in everyday electrical appliances, and we will see alternating current again in Chapter 6.

## 5.11 Inductance

Consider two electrical circuits near each other, as shown schematically in Figure 5.17. The current $I_1$ flowing in circuit (1) creates a magnetic field $\boldsymbol{B} \propto I_1$, whose flux through circuit (2) is

$$\Phi_{B2} \propto I_1.$$

The constant of proportionality is called the *mutual inductance*. That is, we write

$$\Phi_{B2} = M_{21} I_1, \tag{5.21}$$

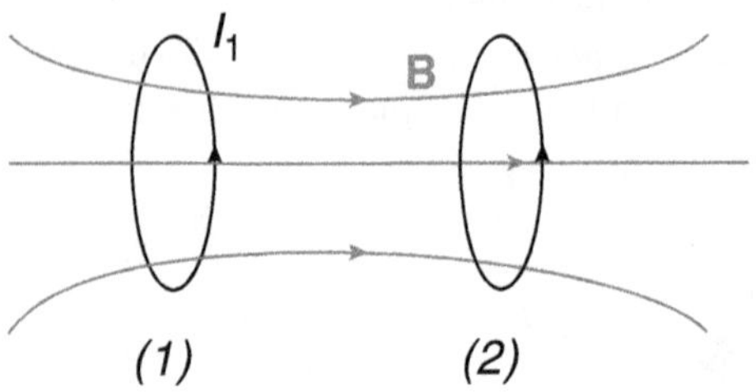


Fig. 5.17 Two circuits near each other.

where $M_{ji}$ is the mutual inductance associated with the field in some circuit $j$ due to a current in circuit $i$. Likewise, one can write

$$\Phi_{B1} = M_{12} I_2, \tag{5.22}$$

which relates the magnetic flux in circuit (1) due to the magnetic field generated by the current $I_2$ in circuit (2). It turns out that in fact the two inductances are equal:

$$M_{21} = M_{12},$$

so that we can use a single symbol $M$ for a given pair of circuits. The value of $M$ depends on the number of loops in each circuit, as well as the geometry of the pair. As an example, let us take the case of two coaxial solenoids, whose geometry is shown in Figure 5.18. If a current $I_b$ flows in the outer solenoid, this creates a magnetic field (see Eq. (5.16))

$$B_b = \frac{\mu_0 I_b N_b}{l}$$

parallel to the axis, where $N_b$ is the number of turns. The flux of this field through a single loop in the inner solenoid is

$$B_b \pi R_a^2,$$

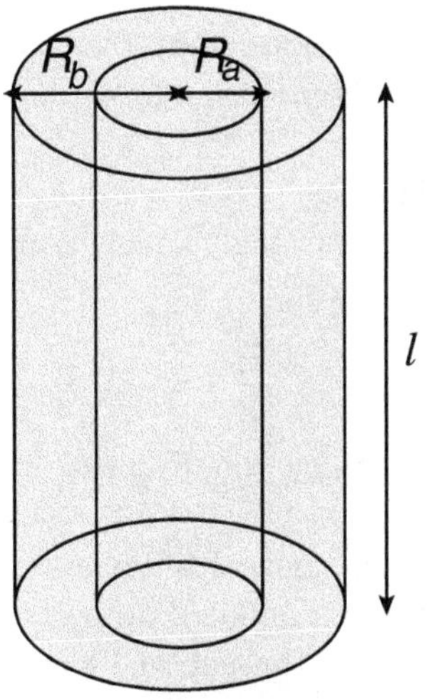


Fig. 5.18 Two coaxial solenoids.

i.e. the field $B_b$ multiplied by the area of the loop. To get the *total* flux through the inner solenoid, we can simply multiply this result by the number of loops:

$$\Phi_{Ba} = N_a B_b \pi R_a^2 = \frac{N_a N_b \mu_0 I_b \pi R_a^2}{l}.$$

By definition one has

$$\Phi_{Ba} = M I_b,$$

and thus we find that the mutual inductance in this case is

$$M = \frac{\mu_0 N_a N_b \pi R_a^2}{l}.$$

Indeed this depends on the number of loops in each circuit, as well as the geomtry of the system. Note that we could also have derived this result by letting the inner solenoid carry a current $I_a$. This would have created a magnetic field

$$B_a = \frac{\mu_0 I_a N_a}{l}$$

parallel to the axis inside the solenoid, and a field which is approximately zero outside. The total flux of this field through the outer solenoid is

$$\Phi_{Bb} = N_b B_a \pi R_a^2 = \frac{\mu_0 N_a N_b \pi R_a^2 I_a}{l} = M I_a,$$

where one must take into account that the field only occupies a cross-sectional area $\pi R_a^2$ of the outer solenoid. We indeed find the same result for $M$ as above.

Given the above definitions, Faraday's law implies that the induced EMF in a circuit (2) due to varying current in another circuit (1) is given by

$$\mathcal{E}_2 = -\frac{d\Phi_{B2}}{dt} = -M\frac{dI_1}{dt},$$

which gives another way to think about mutual inductance. The SI unit of inductance is the *Henry*, where 1 H=1 VsA$^{-1}$ (Volt seconds per amp).

Above, we have seen that mutual inductance is a property of pairs of circuits. A single circuit has a similar property. Its current $I$ creates a magnetic field that has a flux through the circuit. If $I$ varies with time, then the flux will also vary with time, which leads to a *self-induced EMF* in accordance with Faraday's law. Given that the flux is linear in the magnetic field, which itself is linear in the current, one must have $\Phi_B \propto I$, and thus we can write

$$\Phi_B = LI, \tag{5.23}$$

where the constant of proportionality is called the *self-inductance* of the circuit. The self-induced EMF is then

$$\mathcal{E} = -\frac{d\Phi_B}{dt} = -L\frac{dI}{dt}. \tag{5.24}$$

As an example, consider a single solenoid of radius $R$ and length $l$, with $N$ turns and carrying a current $I$. The field inside the solenoid is given by Eq. (5.16), such that the total flux through all of the current loops is

$$\Phi_B = NB\pi R^2 = \frac{\pi R^2 \mu_0 N^2 I}{l} = LI,$$

where the self-inductance in this case is found to be

$$L = \frac{\mu_0 \pi R^2 N^2}{l}. \tag{5.25}$$

If $N$ is large, this self-inductance will be much larger than the self-inductance of any other types of component in a given electrical circuit (including the loops of wire in the circuit itself). Components with nonnegligible self-inductance are called *inductors*, and the circuit symbol is shown in Figure 5.19. By Faraday's law, once a current $I$ is flowing through an inductor, it will induce an EMF (given by Eq. (5.24)) *opposing* the current (if the latter is increasing). An EMF is a potential difference, and a positive EMF opposing the current means that the potential after the inductor must be less than the potential before. Thus, there is a potential difference across the inductor given by

$$\Delta V = -L\frac{dI}{dt}, \tag{5.26}$$

which is negative for $I$ increasing as required. Note that, as for the resistors and capacitors that we have already encountered, it is common to see people define the potential difference in the opposite direction to the current flow,

Fig. 5.19 The circuit symbol for an inductor.

to avoid minus signs. Following Eq. (4.15), we would then say that the potential difference across an inductor is

$$V = L\frac{dI}{dt}. \tag{5.27}$$

Here we have assumed that the current flowing in to the inductor is increasing. If the opposite is true, similar steps will allow us to arrive at Eqs. (5.26) and (5.27), but where the potential difference will now have the opposite sign in each case. We will shortly interpret this physically.

## 5.12 Magnetic Energy

To overcome the induced EMF in an inductor, we must supply energy. This gets stored as potential energy, and we can derive an expression for this as follows. Consider increasing the current through an inductance $L$ from $I = 0$ to $I = I_f$. We saw in Eq. (4.22) that the rate of transfer of energy to an electrical component is given by the current flowing through it, multiplied by the potential difference across it. In the present case this gives

$$|P| = IL\frac{dI}{dt},$$

thus in time $dt$ the change in the internal energy of the inductor is

$$dU = |P|dt = ILdI.$$

The total energy at the end of the process is

$$U = \int_0^{I_f} LdI = \left[\frac{1}{2}LI^2\right]_0^{I_f} = \frac{1}{2}LI_f^2.$$

In other words, an arbitrary inductor through which a current $I$ flows has a potential energy

$$U = \frac{1}{2}LI^2. \tag{5.28}$$

Where does this energy reside? We may think of it as being stored by the magnetic field itself. Consider a solenoid, for example. The inductance is given by Eq. (5.25), where $N$ is the number of turns, $R$ the radius and $l$ the length. From Eq. (5.28), we can then write an expression for the energy per unit volume of the solenoid:

$$\frac{U}{V} = \frac{1}{\pi R^2 l}\frac{1}{2}LI^2 = \frac{1}{2}\frac{\mu_0 N^2 I^2}{l^2}.$$

But the magnetic field inside the solenoid is given by

$$|\boldsymbol{B}| = \mu_0 n I = \frac{\mu_0 N I}{l} \quad \Rightarrow \quad I = \frac{l|\boldsymbol{B}|}{N\mu_0}.$$

We can then write

$$\frac{U}{V} = \frac{1}{2}\frac{\mu_0 N^2}{l^2}\left(\frac{l^2|\boldsymbol{B}|^2}{N^2\mu_0^2}\right) = \frac{|\boldsymbol{B}|^2}{2\mu_0},$$

which shows that the energy density in the inductor can indeed be associated with the magnetic field. This also explains the result we found above, as to why the potential difference can be positive or negative across an inductor: energy can go into or out of the magnetic field, according to how the current is changing.

The final result above turns out to be very general, applicable for *any* magnetic field. We can combine it with our previous result for the energy carried by the electric field (see Section 3.12), so that the total energy density carried by electromagnetic fields is

$$u = \frac{1}{2}\epsilon_0|\boldsymbol{E}|^2 + \frac{1}{2\mu_0}|\boldsymbol{B}|^2. \tag{5.29}$$

We can make similar comments here to what we said about the electric field: the fact that the magnetic field carries energy means that it is a physically real thing, with a life of its own! Furthermore, it will also be able to carry momentum, given that energy and momentum mix with each other when we transform between inertial frames. Note that Faraday's law is the first thing in this book that explicitly relates the electric and magnetic fields. We will carry this further later on by finding the full set of equations that describe all electric and magnetic phenomena. Before doing this, however, we will take a slight detour to examine some of the practical applications of what we have seen so far, making a fuller analysis of electrical circuits.

## Exercises

(1) An electron undergoes circular motion in a magnetic field of 1 T with a speed of $0.5c$, where $c \simeq 3 \times 10^8$ ms$^{-1}$ is the speed of light. Find the radius of the circle, and the period of the motion.

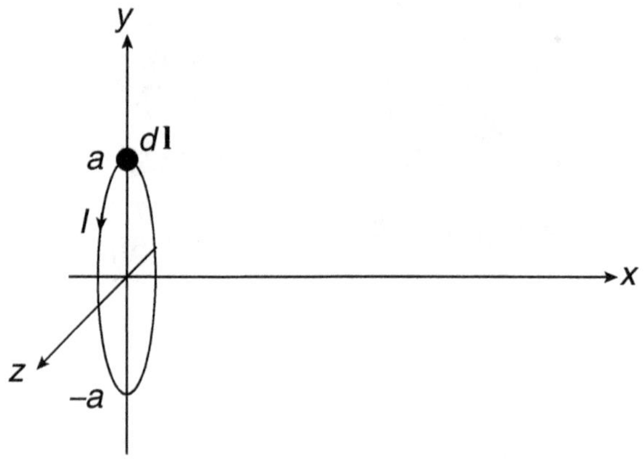


Fig. 5.20 A current-carrying ring, with a segment $\boldsymbol{dl}$ pointing in the direction of the current.

(2) Consider two particles with charge $q_1$ and $q_2$, located at $\boldsymbol{x}_1$ and $\boldsymbol{x}_2$, and moving with velocities $\boldsymbol{v}_1$ and $\boldsymbol{v}_2$. Show that the magnetic force on the charge $q_2$ due to $q_1$ is given by

$$\boldsymbol{F}_{21} = \frac{\mu_0 q_1 q_2}{4\pi} \frac{\boldsymbol{v}_2 \times [\boldsymbol{v}_1 \times (\boldsymbol{x}_2 - \boldsymbol{x}_1)]}{|\boldsymbol{x}_2 - \boldsymbol{x}_1|^3}.$$

(3) Is Newton's third law obeyed by the magnetic force between two charged particles? If it is not true in general, find special cases in which it does apply. Is it a problem if it is not obeyed?

(4) (a) Consider a circular coil of wire of radius $a$ in the $(y, z)$ plane, carrying a current $I$ (see Figure 5.20). Show that the contribution of a segment $\boldsymbol{dl}$ to the $x$-component of the magnetic field at an arbitrary point along the $x$-axis is

$$dB_x = \frac{\mu_0 I a}{4\pi} \frac{dl}{(x^2 + a^2)^{3/2}}.$$

(b) Hence show that the total field at such a point is given by

$$\boldsymbol{B} = \frac{\mu_0 I a^2}{2(x^2 + a^2)^{3/2}} \hat{\boldsymbol{\imath}},$$

where $\hat{\boldsymbol{\imath}}$ is a unit vector in the $x$-direction.

(5) A circular loop of wire of radius 2 cm and carrying current 1 A is placed such that its axis is at 45° to a uniform magnetic field of 1 T. What is the magnitude of the torque experienced by the loop?

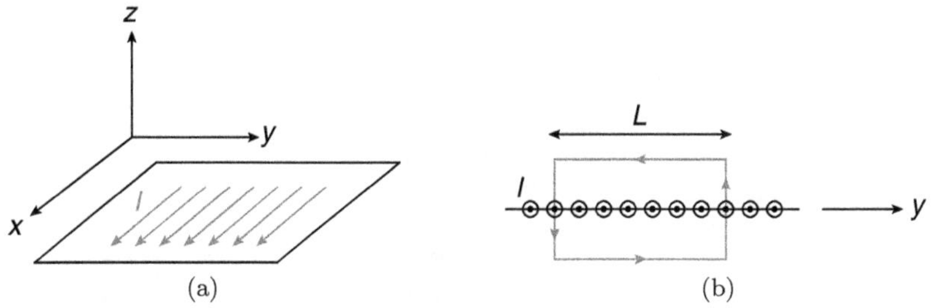


Fig. 5.21 (a) An infinite sheet of current in the $(x, y)$ plane; (b) a view looking down the $x$-axis, such that the current is pointing towards us.

(6) Consider an infinite sheet in the $(x, y)$ plane, carrying a uniform current in the $x$ direction, as shown in Figure 5.21. What does the magnetic field look like away from the sheet? By using Ampère's law with the contour shown in Figure 5.21(b) or otherwise, show that the magnitude of the magnetic field at any point off the sheet is

$$B = \frac{\mu_0 I}{2},$$

where $I$ is the current in the $x$-direction per unit length in the $y$-direction.

# Chapter 6

# A Second Look at Circuits

We have seen some simple electrical circuits already in Chapter 4. In all of those cases, voltages and currents were not changing with time. In most circuits that are used in practical applications, however, this is not the case. It is therefore useful to study the simplest time-dependent circuits, which can be used as building blocks for larger and more complex examples. There are two main types of circuit used in everyday life:

(1) *Direct current (DC)*: In this case, the power source supplies a constant EMF, and the current does not change direction. Examples include any device powered by a simple battery (e.g. a torch).
(2) *Alternating current (AC)*: In this case, the power source supplies a sinusoidally varying voltage, and the current repeatedly switches direction.

Even in the absence of a time-varying source, circuits can be time dependent, due to the nature of the components that they contain. Let us begin, then, by considering some time-dependent DC circuits.

## 6.1 The R-C Circuit

Consider the circuit of Figure 6.1(a), and imagine closing the switch at time $t = 0$, causing a current to flow. Initially, the charge on the capacitor is zero, and the current will cause a charge to build up on the left-hand plate. This in turn will attract or repel charges on the right-hand plate, leading to a build up of equal and opposite charge there. The right-hand plate will in turn attract or repel charges in the wire to the right of the capacitor, so that a current flows around the circuit. Let $I(t)$ be the current at time $t$, and $Q(t)$ the charge on the left-hand capacitor plate. In time $dt$, an additional charge $dQ$ will be deposited on the capacitor, and by the conservation of

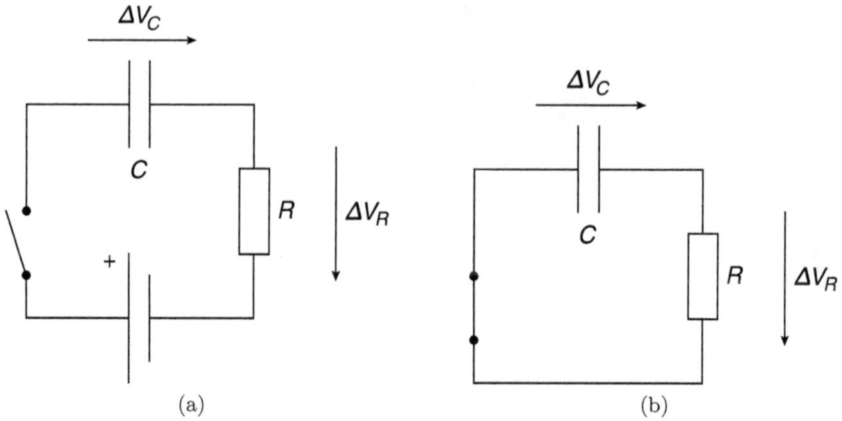


Fig. 6.1 The R-C circuit.

charge, we must have

$$dQ = I(t)dt \tag{6.1}$$

(i.e. the current $I$ is the charge flowing per unit time). We can find an explicit expression for the current by using Kirchhoff's voltage rule, which relates the EMF $\mathcal{E}$ of the power source to the potential differences around the loop (defined in Figure 6.1). Using the convention of Eq. (4.15) where we define $V_i$ to be the potential difference measured against the current flow, we can write Kirchoff's law in this case as

$$\begin{aligned}\mathcal{E} &= V_R + V_C \\ &= I(t)R + \frac{Q(t)}{C},\end{aligned} \tag{6.2}$$

where in the second line we have used Ohm's law, and the definition of capacitance, in the first and second terms, respectively. We thus have

$$I = \frac{\mathcal{E}}{R} - \frac{Q}{RC} = -\frac{1}{RC}\left(Q - C\mathcal{E}\right).$$

Substituting this into Eq. (6.1), we have

$$dQ = -\frac{1}{RC}(Q - C\mathcal{E})dt \quad \Rightarrow \quad \frac{dQ}{Q - C\mathcal{E}} = -\frac{dt}{RC}.$$

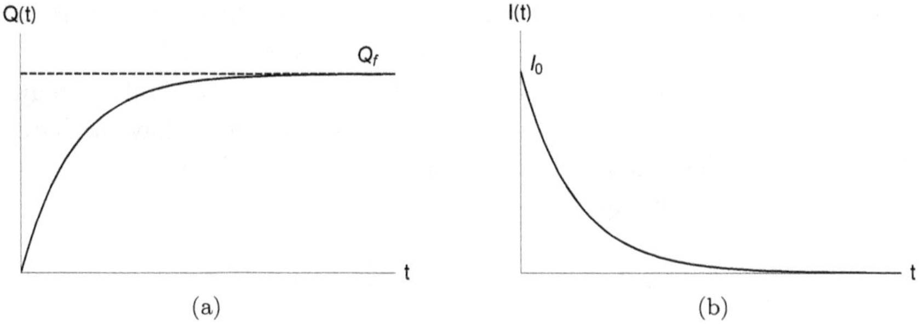


Fig. 6.2 (a) The charge on a charging capacitor; (b) the current flowing in a charging capacitor.

We can integrate this equation from time $t = 0$ to some time $t$:

$$\int_0^{Q(t)} \frac{dQ}{Q - C\mathcal{E}} = \int_0^t dt' \left(-\frac{1}{RC}\right),$$

which can be rearranged to give

$$Q(t) = C\mathcal{E}[1 - e^{-t/RC}].$$

We can neaten this result slightly by noting that as $t \to \infty$, $Q(t) \to C\mathcal{E}$, which is a constant. We can simply label this by $Q_f$, where the $f$ stands for "final value of the charge". Then we can write

$$Q(t) = Q_f[1 - e^{-t/RC}]. \tag{6.3}$$

A plot of this function is shown in Figure 6.2: the charge increases from zero at $t = 0$, before saturating at a constant value some time later. The current will be given by

$$I(t) = \frac{dQ}{dt} = \frac{\mathcal{E}}{R} e^{-t/RC}.$$

But from Eq. (6.2) and the fact that $Q(0) = 0$, we can write

$$I(0) = \frac{\mathcal{E}}{R} \equiv I_0,$$

so that

$$I(t) = I_0 e^{-t/RC}. \tag{6.4}$$

A plot of this dependence is shown in Figure 6.2(b): the current starts at a nonzero value, before exponentially decaying to zero.

As well as a charging capacitor, we can also consider a similar circuit, shown in Figure 6.1(b), in which a capacitor is assumed to have a charge $Q_0$ at $t = 0$, at which time the switch is opened. In the absence of a source, Kirchhoff's voltage rule now becomes

$$\Delta V_C + \Delta V_R = 0 \quad \Rightarrow \quad \frac{Q(t)}{C} + I(t)R = 0.$$

The charge on the capacitor then satisfies

$$dQ = Idt = -\frac{Q}{RC}dt \quad \Rightarrow \quad \ln\left(\frac{Q}{Q_0}\right) = -\frac{t}{RC}.$$

Finally, we obtain

$$Q = Q_0 e^{-t/RC}. \tag{6.5}$$

The current is given by

$$I = \frac{dQ}{dt} = -\frac{Q_0}{RC}e^{-t/RC}.$$

Again defining $I(0) \equiv I_0$, we can rewrite this as

$$I = I_0 e^{-t/RC}, \quad I_0 = -\frac{Q_0}{RC}. \tag{6.6}$$

Example plots of the charge and current are shown in Figure 6.3. The charge starts at a non-zero value, before exponentially decaying to zero.

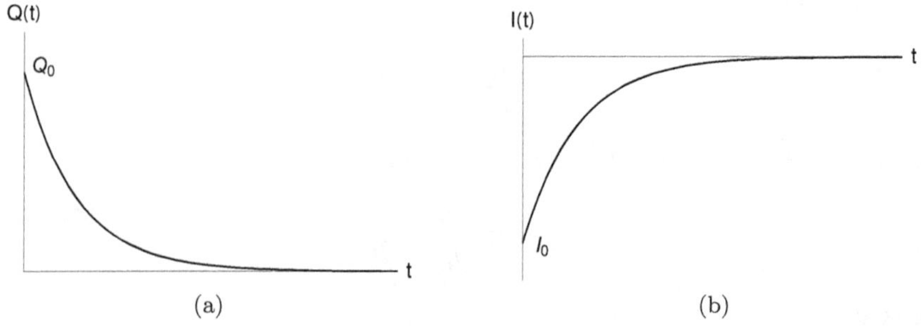


Fig. 6.3 (a) The charge on a discharging capacitor; (b) the current flowing in a discharging capacitor.

The ensuing current is *negative*, as charge leaves the capacitor in the opposite direction to the current that charged it up in the first place. It reaches zero once the capacitor is fully discharged. Note that in both the charging and discharging cases, the quantity

$$\tau = RC \tag{6.7}$$

measures how long it takes the system to reach a steady state. That is, the capacitor has roughly fully (dis)charged for $t > \tau$. This quantity is known as the *time constant*, or *relaxation time*, and we will see similar quantities occuring in other types of circuit.

## 6.2 The R-L Circuit

In this section, we begin by considering the circuit shown in Figure 6.4(a), and containing a single inductor and resistor. Let the switch be closed at time $t = 0$. Kirchhoff's voltage rule then gives

$$\mathcal{E} = V_R + V_L = IR + L\frac{dI}{dt},$$

where $\mathcal{E}$ is the EMF of the source, and we have used Ohm's law and Eq. (5.26) in the second equality. Rearranging, the current obeys the

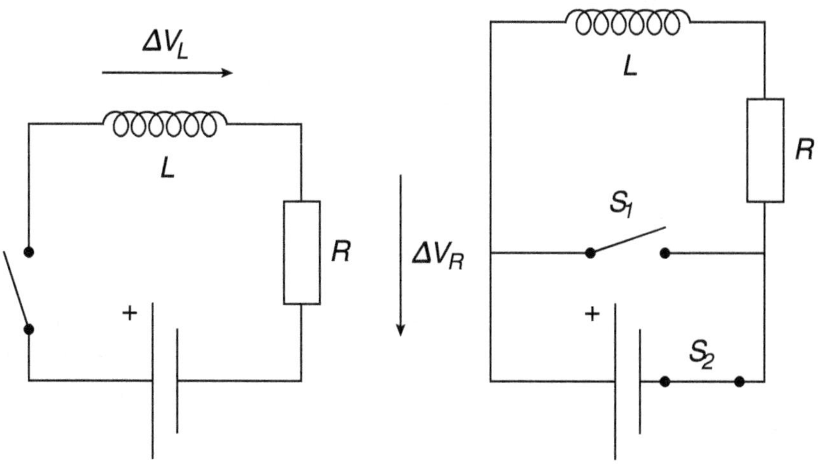


(a) (b)

Fig. 6.4 The R-L circuit.

differential equation

$$\frac{dI}{dt} = -\frac{R}{L}\left(I - \frac{\mathcal{E}}{R}\right) \quad \Rightarrow \quad \int_0^{I(t)} \frac{dI}{I - \mathcal{E}/R} = -\int_0^t \frac{R}{L} dt'.$$

Integrating both sides and rearranging, we obtain

$$I = \frac{\mathcal{E}}{R}[1 - e^{-Rt/L}]. \tag{6.8}$$

The functional form of this looks similar to Figure 6.2(a): the current rises from 0 to the value obtained if the inductor is absent. The circuit is often used in practice to avoid potentially damaging spikes of current (e.g. due to power surges) in larger circuits: the current is forced to rise gradually, instead of suddenly jumping to a higher value.

As well as growing, current can also *decay* in R-L circuits. Consider the circuit of Figure 6.4(b), where we assume that a current $I_0$ is intially flowing, having reached its stable value. Then, at time $t = 0$, we open switch $S_2$ and close switch $S_1$. Current now flows around the upper loop, rather than the outer loop containing the source. We can again find the current using Kirchhoff's voltage law, which now takes the form

$$V_R + V_L = 0 \quad \Rightarrow \quad IR + L\frac{dI}{dt} = 0.$$

Rearranging gives

$$\int_{I_0}^{I(t)} \frac{dI}{I} = -\int_0^{t'} \frac{R}{L} dt' \quad \Rightarrow \quad \ln\left(\frac{I}{I_0}\right) = -\frac{Rt}{L}.$$

Finally, we get

$$I = I_0 e^{-tR/L}. \tag{6.9}$$

The functional form looks similar to Figure 6.3(a). That is, the current starts at some non-zero value, before exponentially decaying to zero. As for the charge in the R-L circuit of the previous section, we can talk about a *time constant* $\tau$, such that the current has more or less decayed for $t > \tau$. In this case the relevant quantity is

$$\tau = \frac{L}{R}. \tag{6.10}$$

## 6.3 The L-C Circuit

The circuits in the previous two sections show exponential behaviour for charges and currents. This is not the only possibility, and in particular it is also possible to produce *oscillatory* behaviour. Consider the circuit of Figure 6.5, containing an inductor and a capacitor. Let us assume that the capacitor is initially fully charged, and we may then consider moving the switch at time $t = 0$ from position $A$ to position $B$. Kirchhoff's voltage rule dictates that the potential differences across the inductor and capacitor satisfy

$$V_L + V_C = 0 \quad \Rightarrow \quad L\frac{dI}{dt} + \frac{Q}{C} = 0.$$

But we also know that

$$I = \frac{dQ}{dt}, \tag{6.11}$$

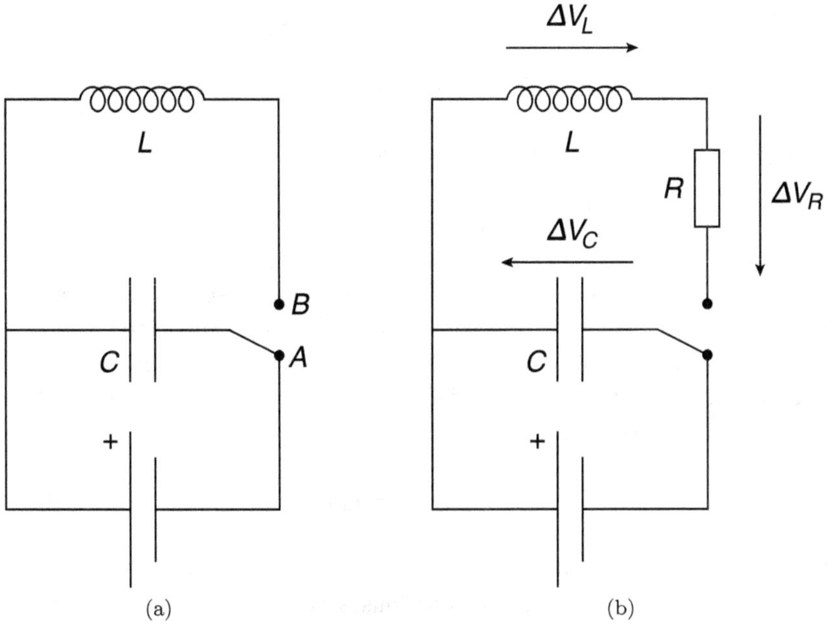


Fig. 6.5 The L-C circuit.

where $Q$ is the charge on the left-hand (positive) capacitor plate. This charge thus satisfies the second-order differential equation

$$L\frac{d^2Q}{dt^2} + \frac{Q}{C} = 0.$$

or, upon rearranging,

$$\frac{d^2Q}{dt^2} = -\frac{1}{LC}Q. \tag{6.12}$$

This equation has the form

$$\frac{d^2Q}{dt^2} = -\omega^2 Q, \quad \omega^2 = \frac{1}{LC},$$

which is the equation of *simple harmonic motion* (*SHM*), that is often encountered in mechanical systems. The general solution is

$$Q = Q_0 \cos(\omega t + \phi), \tag{6.13}$$

for some constants $Q_0$ and $\phi$. We can check this as follows. Differentiating Eq. (6.13) once gives

$$\frac{dQ}{dt} = -\omega Q_0 \sin(\omega t + \phi),$$

so that differentiating a second time yields

$$\frac{d^2Q}{dt^2} = -\omega^2 Q_0 \cos(\omega t + \phi) = -\omega^2 Q,$$

as required. Furthermore, there are two arbitrary constants ($Q_0$ and $\phi$) in Eq. (6.13), as expected for a second-order differential equation. The form of Eq. (6.13) is shown, for a particular choice of constants, in Figure 6.6, and we can see that it indeed oscillates between $Q_0$ and $-Q_0$. The period of the oscillation is given by

$$T = \frac{2\pi}{\omega} = 2\pi\sqrt{LC}. \tag{6.14}$$

To understand why this is happening, we can apply what we learnt about capacitors and inductors in the previous sections. Consider the start of each oscillation, at which time the capacitor is fully charged with charge $Q_0$ on its left-hand plate, so that (as we found in Section 6.1) the current

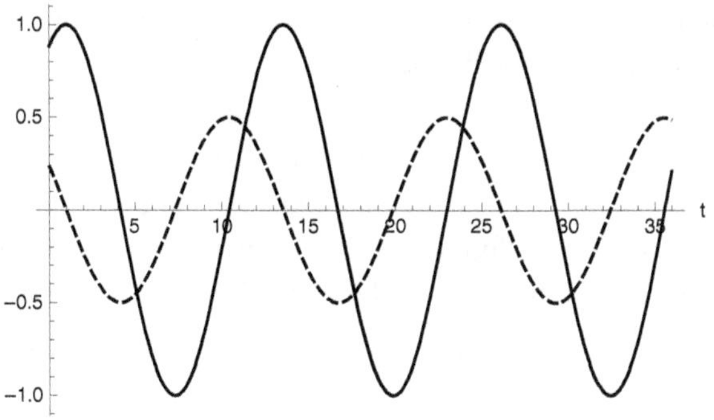


Fig. 6.6 The charge (solid) and current (dashed) in the L-C circuit in SI units, where $Q_0 = 1$ C, $\phi = -0.5$ and $\omega = 0.5$ s$^{-1}$.

will be zero. The capacitor will then discharge, causing a negative current. However, the current cannot rise instantly, and is instead held back by the presence of the inductor — as we found in Section 6.2. The current thus rises gradually, reaching a maximum magnitude when the capacitor is fully discharged, so that the charge $Q$ is zero. The current will persist, and charge up the capacitor with the *opposite* polarity to its initial configuration. The capacitor will be fully charged when the charge on its left-hand plate is $-Q_0$, during which process the current has gradually decreased in magnitude to zero. At this point, the capacitor will discharge again, but causing a *positive current*, so that the charge falls to zero but then rises again, returning the system back to its initial state. The process then repeats over and over again, with charge sloshing from one plate of the capacitor to the other.

We can obtain an explicit expression for the current:

$$I = \frac{dQ}{dt} = -\omega Q_0 \sin(\omega t + \phi), \tag{6.15}$$

whose form is shown in Figure 6.6, in agreement with the qualitative discussion above. The current is a quarter of a cycle out of phase with the charge, and this is analogous to the fact that, in mechanical examples of SHM, the velocity is similarly out of phase with the displacement.

In mechanical systems, the energy in SHM oscillates between potential and kinetic energy. We can also see what is going on in our electrical system. The total energy at any given time will be given by the sum of the energies in the capacitor and inductor:

$$E_{\text{tot.}} = E_C + E_L.$$

For the capacitor, Eq. (3.30) implies that the (potential) energy associated with the charge in the capacitor is

$$E_C = \frac{1}{2}QV,$$

where $V$ is the potential difference on the positive plate, and we may choose $V = 0$ on the negative plate. From the definition of capacitance (Eq. (4.2)), we can rewrite this as

$$E_C = \frac{1}{2}\frac{Q^2}{C}.$$

The energy in the inductor is given by Eq. (5.28), so that we find

$$\begin{aligned} E_{\text{tot.}} &= \frac{1}{2}\frac{Q^2}{C} + \frac{1}{2}LI^2 \\ &= \frac{1}{2C}Q_0^2\cos^2(\omega t + \phi) + \frac{L}{2}\omega^2 Q_0^2 \sin^2(\omega t + \phi), \end{aligned}$$

where we have substituted the results of Eqs. (6.13) and (6.15). We may now use the fact that $\omega^2 = 1/(LC)$ to get

$$\begin{aligned} E_{\text{tot.}} &= \frac{Q_0^2}{2C}\left[\cos^2(\omega t + \phi) + \sin^2(\omega t + \phi)\right] \\ &= \frac{Q_0^2}{2C}, \end{aligned} \tag{6.16}$$

where we have used the trigonometric identity

$$\cos^2(z) + \sin^2(z) = 1,$$

valid for an arbitrary (complex) $z$. The final result of Eq. (6.16) is constant for a given L-C circuit, as expected: energy is conserved in simple harmonic motion. As time progresses, the energy in the system oscillates between the electric field in the capacitor, and the magnetic field in the inductor.

## 6.4 The L-R-C Circuit

We can also add a resistor to the L-C circuit, to make the circuit shown in Figure 6.5(b). Again, let us take the capacitor to be fully charged at time $t = 0$, and then flick the switch between position $A$ and position $B$. Now

Kirchhoff's voltage rule applied to the upper loop implies

$$V_C + V_R + V_L = \frac{Q}{C} + IR + L\frac{dI}{dt} = 0.$$

Again substituting Eq. (6.11), the differential equation for the charge on the left-hand capacitor plate gets modified to

$$\frac{d^2Q}{dt^2} + \frac{R}{L}\frac{dQ}{dt} + \frac{Q}{LC} = 0. \tag{6.17}$$

The additional term is linear in the first derivative of the charge, which is analogous to the velocity in a mechanical system. Thus, this is like adding a damping term in conventional SHM, and to solve the equation we can try solutions of the form

$$Q = Ae^{\alpha t}$$

which, upon substitution into Eq. (6.17), yields

$$A\left[\alpha^2 + \frac{R\alpha}{L} + \frac{1}{LC}\right]e^{\alpha t} = 0.$$

We have indeed found a solution, provided

$$\alpha^2 + \frac{R\alpha}{L} + \frac{1}{LC} = 0.$$

In fact, this has two solutions in general, with

$$\alpha = -\frac{R}{2L} \pm \frac{1}{2}\left(\frac{R^2}{L^2} - \frac{4}{LC}\right)^{1/2}.$$

Consider, for example, the case where

$$R^2 < \frac{4L}{C}.$$

The argument of the square root is then negative, so that we can write

$$\alpha = -\frac{R}{2L} \pm i\left(\frac{1}{LC} - \frac{R^2}{4L^2}\right)^{1/2},$$

where now the square root is manifestly real. The full solution can now be written

$$Q = e^{-Rt/2L}\left[Ae^{i\omega t} + Be^{-i\omega t}\right], \tag{6.18}$$

where, in contrast to the previous section, we now have

$$\omega^2 = \frac{1}{LC} - \frac{R^2}{4L^2}.$$

Using Euler's theorem

$$e^{iz} = \cos(z) + i\sin(z),$$

we see that Eq. (6.18) is indeed oscillatory with angular frequency $\omega$. Furthermore, the overall amplitude is an exponentially decaying function, so that the oscillations die away as $t \to \infty$, as shown in Figure 6.7. The constants $A$ and $B$ in Eq. (6.18) are arbitrary, apart from the fact that, in any given circuit, they must be such as to ensure that the charge $Q$ is real. We can then rewrite the general solution as

$$Q = Q_0 e^{-Rt/2L} \cos(\omega t + \phi). \tag{6.19}$$

A plot of this, for some example choices of the various parameters, is shown in Figure 6.7. We have here focused on the case in which the damping is light, so that many oscillations occur before the amplitude dies away. Other solutions exist, namely for *heavy damping*, in which no oscillations occur at all, and *critical damping*, in which a single oscillation effectively occurs before the amplitude decays to zero.

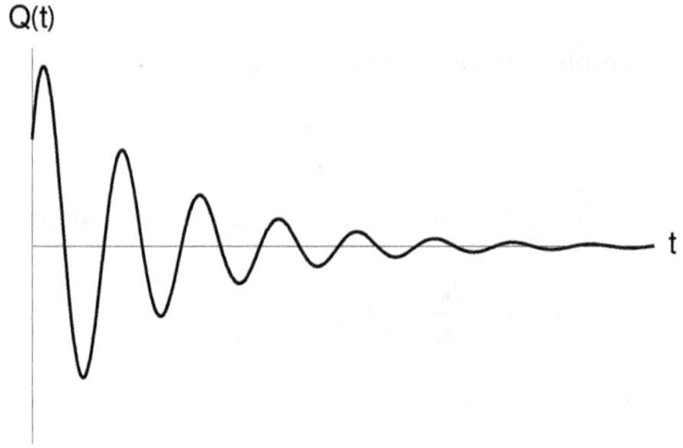


Fig. 6.7 Example solution for the charge on the capacitor in an L-R-C circuit.

## 6.5 Alternating Current

All of the previous examples of circuits that we have analysed have had either no EMF source, or a source providing a constant source of potential difference. This is known as *DC*, and is not how most everyday electrical appliances work. Instead, they use *AC*, for which the supplied voltage has an oscillatory form:

$$V = V_0 \cos(\omega t), \tag{6.20}$$

with some constant amplitude $V_0$, and angular frequency $\omega$. The main reason for using AC is that is much easier to transport AC power over large distances (e.g. using transmission lines). Furthermore, it is straightforward to step-up and step-down AC voltages, once the electricity has been transported from a power station to a home or office.

AC circuits will have an oscillating current as well as voltage. However, the two quantities will not necessarily be in phase, due to the various electrical components that may be present. To analyse this, it is highly convenient to introduce the idea of *complex voltage*. That is, we may write the voltage in an AC circuit as

$$V = V_0 e^{i\omega t}, \tag{6.21}$$

so that the actual (measured) voltage is

$$\text{Re}(V) = \text{Re}\left[V_0 \left(\cos(\omega t) + i\sin(\omega t)\right)\right] = V_0 \cos(\omega t). \tag{6.22}$$

As Eq. (6.21) makes clear, complex numbers have a phase as well as a magnitude, and thus writing circuit properties in terms of complex numbers proves to be a highly efficient way to keep track of phase *differences* between, e.g. voltage and current. We can always convert back to measurable quantities by taking real parts.

Following what we have done for voltage, a general current can be written as a complex number:

$$I = I_0 e^{i(\omega t + \phi)}, \tag{6.23}$$

where we have introduced a *phase shift* $\phi$ with respect to the voltage in general. The actual current in the circuit is then

$$\text{Re}(I) = I_0 \cos(\omega t + \phi). \tag{6.24}$$

The above expressions imply that we can write

$$V = IZ,$$

which defines the *complex impedance*

$$Z = \frac{V}{I} = \frac{V_0}{I_0} e^{-i\phi}. \tag{6.25}$$

In the first equality, $V$ and $I$ are complex voltage and current, respectively, and we see in the second equality that the complex impedance measures two things:

(i) The magnitude

$$|Z| = \frac{V_0}{I_0}$$

relates the amplitudes of the voltage and current. This is also called the *reactance* $X$ in electronics textbooks.

(ii) The argument $\arg(z)$ measures the *phase difference* between the voltage and current.

To see how the voltage across various components is related to the current through them, let us examine some specific examples of the complex impedance.

**1. A resistor**

At any time $t$, the current and voltage across a resistance $R$ are related by Ohm's law, Eq. (4.13). Taking this to apply to the complex quantities, we find that the complex impedance for a resistor is purely real:

$$Z_R = R = X_R, \quad \arg(Z_R) = 0. \tag{6.26}$$

That is, the impedance and reactance are both the same as the resistance $R$, and there is no phase shift between $V$ and $I$.

**2. A capacitor**

The charge on a capacitor is given by Eq. (4.2). By conservation of charge, a current $I$ deposits a charge

$$dQ = Idt$$

on a capacitor in time $dt$, so that one has

$$I = \frac{dQ}{dt} = C\frac{dV}{dt}. \tag{6.27}$$

Taking the voltage across the capacitor to have its complex AC form of Eq. (6.21), we find

$$\begin{aligned} I &= i\omega C V_0 e^{i\omega t} \\ &= \omega C V_0 e^{i\omega t} e^{i\pi/2} \\ &= \omega C e^{i\pi/2} V. \end{aligned}$$

We thus find that the complex impedance for a capacitor is

$$Z_C = \frac{1}{\omega C} e^{-i\pi/2}. \tag{6.28}$$

This has a reactance

$$X_C = |Z_C| = \frac{1}{\omega C},$$

that, by comparison with the previous example, will look like a resistance. The argument of the impedance is non-zero:

$$\arg(Z_C) = -\frac{\pi}{2}, \tag{6.29}$$

so that the current is a quarter of a cycle ahead of the voltage, as shown in Figure 6.8.

### 3. An inductor

The voltage across an inductor satisfies (from Eq. (5.26))

$$V = L\frac{dI}{dt},$$

Taking the voltage to be that of an AC circuit, we then have

$$I = \frac{1}{L}\int dt V_0 e^{i\omega t} = -\frac{iV_0}{\omega L} e^{i\omega t} = \frac{e^{-i\pi/2}}{\omega L} V.$$

Thus, the complex impedance of an inductor is given by

$$Z_L = \omega L e^{i\pi/2}. \tag{6.30}$$

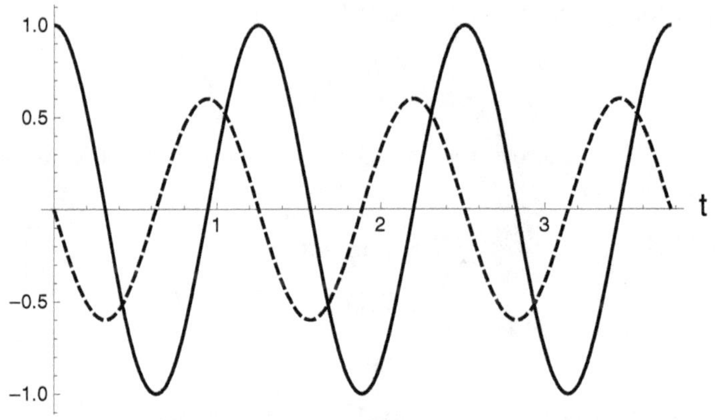


Fig. 6.8 The voltage across a capacitor in an AC circuit (solid), and the current flowing out of it (dashed).

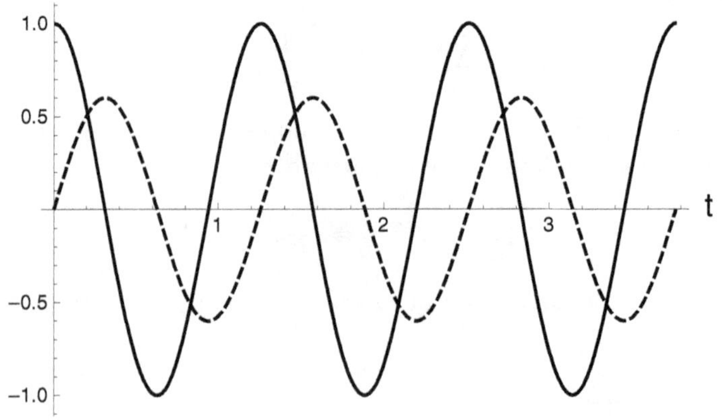


Fig. 6.9 The voltage across an inductor in an AC circuit (solid), and the current through it (dashed).

The reactance is

$$X_L = |Z_L| = \omega L, \tag{6.31}$$

and there is now a phase difference

$$\arg(Z_L) = \frac{\pi}{2}, \tag{6.32}$$

such that the current is a quarter of a cycle behind the voltage, as shown in Figure 6.9.

## 6.6 The AC L-R-C Circuit

Consider the circuit shown in Figure 6.10, consisting of a resistor, inductor and capacitor in series, together an AC source, that supplies a complex EMF

$$\mathcal{E} = V_0 e^{i\omega t},$$

i.e. such that the actual EMF is the real part of this. Applying Kirchhoff's voltage law around the circuit gives

$$\mathcal{E} = V_C + V_L + V_R,$$

and applying results from the previous section, this becomes

$$\frac{1}{C}\int I dt + L\frac{dI}{dt} + IR = V_0 e^{i\omega t}. \tag{6.33}$$

We saw before that, without the source, this leads to a damped, oscillatory current (n.b. earlier we wrote this equation in terms of the charge rather than the current). With the source included, the system is a *driven simple harmonic oscillator* where, by analogy with mechanical systems, the non-zero right-hand side acts as a "driving force", which oscillates with angular frequency $\omega$. To solve Eq. (6.33), we can try a solution oscillating with the

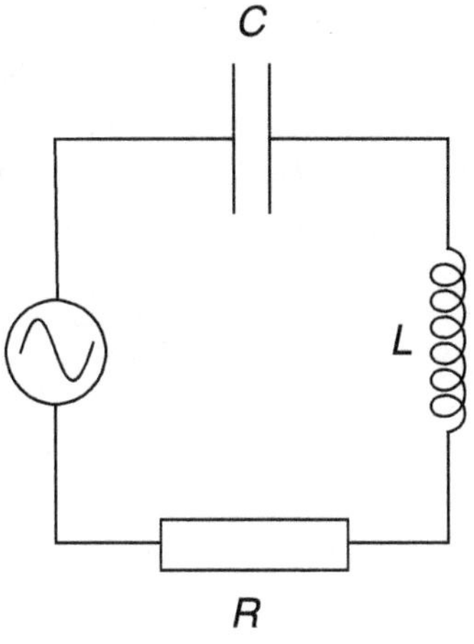


Fig. 6.10 An L-R-C circuit, with an alternating current source.

same frequency as the driving force:

$$I = I_0 e^{i\omega t},$$

and substituting this into Eq. (6.33) yields

$$\left[-\frac{i}{\omega C} + i\omega L + R\right] I_0 e^{i\omega t} = V_0 e^{i\omega t}.$$

Thus, we have indeed found a solution, provided

$$V = ZI,$$

with

$$Z = R + i\left(\omega L - \frac{1}{\omega C}\right). \tag{6.34}$$

This has a nice interpretation. The ratio of (complex) voltage to current is the complex impedance of the whole circuit. Our solution can be identified as the current obtained from the source voltage due to this complex impedance, and the total impedance is just the sum of the individual complex impedances that we found for each component in the previous section. Note, however, that this is not the complete solution for the current in the circuit. According to standard lore on linear ordinary differential equations, we could add to our solution for $I$ a solution of the *homogeneous* equation consisting of Eq. (6.33) with no driving force on the right-hand side. We know already, however, what solutions of the homogeneous equation correspond to: they are *transient* damped oscillations, that die away provided we wait long enough. Thus, we will not consider them further.

For a given amplitude of the source voltage $V_0$, we can maximise the amplitude of the current, $I_0 = |I|$, by tuning the angular frequency $\omega$ of the source. From the above results, we have

$$I_0 = \frac{V_0}{|Z|} = \frac{V_0}{\left|R + i\left(\omega L - \frac{1}{\omega C}\right)\right|} = \frac{V_0}{\sqrt{R^2 + \left(\omega L - \frac{1}{\omega C}\right)^2}}, \tag{6.35}$$

and a plot of this as a function of $\omega$ is shown in Figure 6.11. It displays a sharp peak at a particular value of $\omega$, and from Eq. (6.35) we can see that the function will be maximised when the contents of the bracket in the denominator vanishes, i.e. when

$$\omega^2 \equiv \omega_0^2 = \frac{1}{LC}. \tag{6.36}$$

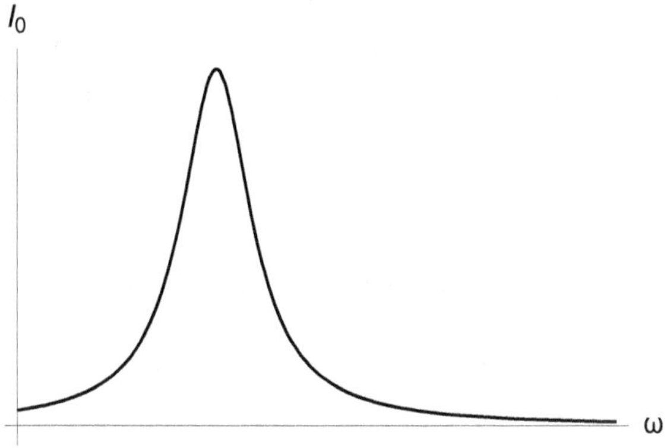


Fig. 6.11 Amplitude of the current in an AC L-R-C circuit, as a function of the angular frequency $\omega$ of the source.

This phenomenon is called *resonance*, and is a feature of any simple harmonic oscillator driven by a periodic driving force, namely that the amplitude of the oscillation reaches a maximum for a particular *resonant frequency*. From the results of Section 6.3, we can see that the resonant frequency of the electrical circuit corresponds to the frequency of the undamped, non-driven oscillator. Other examples of resonance also display this behaviour: oscillatory systems in general have a particular set of associated frequencies, that will act as resonant frequencies when a periodic driving force is used to excite the system.

There are many applications of resonance in electrical circuits. An example from everyday life is tuning a radio. The circuit involved in that case will be designed to exhibit resonance at the frequency corresponding to a particular radio station, so that detection of the signal is maximised.

## Exercises

(1) A DC R-C circuit contains a resistor of 1 $\Omega$, and a capacitor of 4 $\mu$F. If the circuit is broken at time $t = 0$, roughly how long does it take for the capacitor to discharge?

(2) (a) Consider two inductors in series. By considering the potential difference across the inductors, show that the system behaves like a *single* inductor, with inductance

$$L = L_1 + L_2.$$

(b) Assume that both inductors are cylindrical solenoids with the same radius $R$, and number of turns per unit length (although the total number of turns may be different). The inductance of such a solenoid is given by

$$L = \frac{\mu_0 \pi R^2 N^2}{l}.$$

Show that this result is consistent with the result of part (a).

(3) Show that the rules for combining capacitors in series and in parallel follow from assuming that one combines complex impedances similarly to resistances: that is,

$$Z = Z_1 + Z_2,$$

for impedances in series, and

$$\frac{1}{Z} = \frac{1}{Z_1} + \frac{1}{Z_2},$$

for impedances in parallel.

# Chapter 7

# Maxwell's Equations

## 7.1 The Displacement Current

We have now seen almost all the physical laws that are needed to describe the entirety of electromagnetism. To recap, we have seen Gauss' Law for electric and magnetic fields — Eqs. (3.18) and (5.4), respectively — together with Ampère's law for magnetic fields (Eq. (5.15)), and Faraday's law (Eq. (5.18)). The first two of these equations involve the flux of the electric and magnetic fields through arbitrary closed surfaces, and the second two equations involve the line integral of the fields around arbitrary closed loops. It turns out that this will indeed be all that is needed to fully determine the behaviour of electric/magnetic fields in all circumstances. However, there is a rather subtle problem with Ampère's law as we have currently formulated it.

The right-hand side of Ampère's law contains the current through a surface spanned by the closed contour $C$, and we are meant to be able to choose *any* such surface. It is straightforward to see that this fails inside a capacitor. Take, for example, a parallel plate capacitor, with a closed curve $C$ enclosing the incoming wire. We may then consider the surface $S_1$ bounded by the curve, as shown in Figure 7.1(a). Ampère's law applied to the surface $S_1$ gives

$$\oint_C \boldsymbol{B} \cdot d\boldsymbol{l} = \mu_0 I, \tag{7.1}$$

given that all of the incoming current $I$ crosses the surface. We may instead consider a surface that reaches between the capacitor plates, such as the surface $S_2$ shown in Figure 7.1(b). No current crosses this surface, and thus

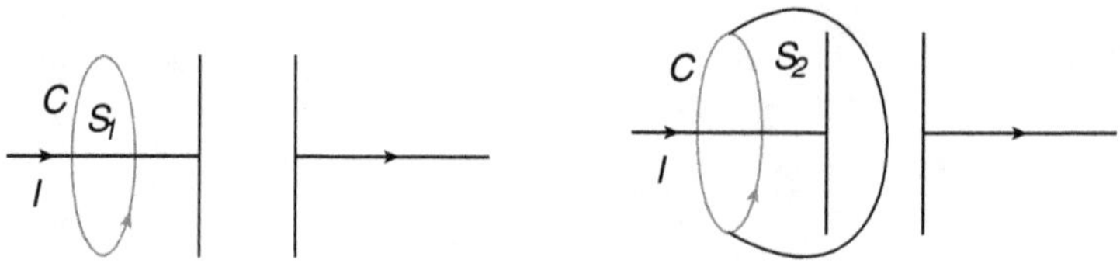


Fig. 7.1 (a) A closed contour $C$ around a wire incoming to a capacitor, where $S_1$ is the minimal surface bounded by the contour; (b) alternative surface $S_2$ that reaches between the capacitor plates.

Ampère's law gives

$$\oint_C \boldsymbol{B} \cdot d\boldsymbol{l} = 0, \tag{7.2}$$

in clear contradiction with Eq. (7.1)! We could simply take a more careful definition of Ampère's law, namely that the current enclosed has to be that crossing the minimal surface $S_1$ bounded by the contour. However, this seems an odd thing to do, given that Ampère's law also works for many other surfaces, that are not the minimal surface $S_1$. Furthermore, the basic reason that Ampère's law is failing is that there is no conventional current flowing between the capacitor plates. The mechanism for current flowing in a circuit containing a capacitor is that the positive charge on one plate attracts or repels charges on the other plate, creating a net negative charge. This then causes charges on the other side of the capacitor to move around the circuit, and thus it makes sense to define a current as flowing "through" the capacitor, even though the *conventional* current (i.e. the rate of flow of charge) $I$ is zero. If we could define some sort of current that was non-zero inside the capacitor, we could patch up Ampère's law. Furthermore, we could apply Kirchoff's current law everywhere in the circuit, which is a sensible thing to be able to do given that charge is conserved everywhere.

We can formalise the above discussion by postulating that there is an *extra term* on the right-hand side of Ampère's law. First, note that the charge on the left-hand capacitor plate is given by

$$Q = C|\Delta V|,$$

where

$$C = \frac{\epsilon_0 A}{d}, \quad |\Delta V| = |\boldsymbol{E}|d$$

are the capacitance and potential difference across the capacitor respectively (see Section 4.1), $A$ the area of each plate, and $d$ the separation between the plates. Also, $\boldsymbol{E}$ is the electric field between the plates, and we may use these results to write

$$Q = \epsilon_0 \Phi_E, \tag{7.3}$$

where $\Phi_E = EA$ is the flux of $\boldsymbol{E}$ through the surface $S_2$ (given there are no charges in between the capacitor plates, the flux through $S_2$ must be the same as through a surface that just covers the left-hand plate of the capacitor, with area $A$). In time $dt$, a charge

$$dQ = I dt$$

is deposited on the capacitor, so that the conventional current is given by

$$I = \frac{dQ}{dt}.$$

It then follows from Eq. (7.3) that we can regard current as being conserved through the capacitor by defining the so-called *displacement current*

$$I_d = \epsilon_0 \frac{d\Phi_E}{dt}.$$

We can then add this to the current $I$ on the right-hand side of Ampère's law, so that the latter is replaced by

$$\oint_C \boldsymbol{B} \cdot d\boldsymbol{l} = \mu_0 I + \mu_0 \epsilon_0 \frac{d\Phi_E}{dt}. \tag{7.4}$$

The displacement current was first introduced by Maxwell, who was trying to collect all of the equations of electricity and magnetism into a single, consistent mathematical framework. Thus, for want of a better name, I will call Eq. (7.4) *Maxwell's Ampère's Law.* In the above example, there is a conventional current $I$ only outside the capacitor, and a displacement current only inside the capacitor. These have the same magnitude, so that indeed

$$\oint_C \boldsymbol{B} \cdot d\boldsymbol{l}$$

gives the same result, for *any* surface spanning the closed contour $C$. More generally, Maxwell's Ampère's law applies in all situations, for any combination of conventional and displacement currents.

## 7.2 Maxwell's Equations

We can now collect all of the fundamental laws we have seen into a single set of four equations:

$$\oiint_S \boldsymbol{E} \cdot d\boldsymbol{S} = \frac{Q}{\epsilon_0}; \tag{7.5}$$

$$\oiint_S \boldsymbol{B} \cdot d\boldsymbol{S} = 0; \tag{7.6}$$

$$\oint_C \boldsymbol{E} \cdot d\boldsymbol{l} = -\frac{d\Phi_B}{dt}; \tag{7.7}$$

$$\oint_C \boldsymbol{B} \cdot d\boldsymbol{l} = \mu_0 I + \mu_0 \epsilon_0 \frac{d\Phi_E}{dt}. \tag{7.8}$$

These are the famous *Maxwell equations*, and tell us, in full generality, how electric and magnetic fields are generated from currents and charges. They do not, however, tell us how charged particles behave in response to the fields. For this, we must add the Lorentz force law, which says that the total electromagnetic force on a particle of charge $q$ is

$$\boldsymbol{F} = q(\boldsymbol{E} + \boldsymbol{v} \times \boldsymbol{B}). \tag{7.9}$$

But that is everything — we now have the complete theory of electromagnetism! What's more, this takes less than half a page to write down, and yet is able to describe a truly amazing number of different phenomena, from everyday life and beyond. To try to get you to appreciate the significance of this, think of what you do in a typical day. You might first notice that the sky is blue in the morning and afternoon, but red at dusk. Or it is cloudy, in which case you see a white or grey sky. You may use various home appliances in the morning to help get yourself (or your breakfast!) ready. You might check the news, or communicate with friends and family, using your phone or computer. It is possible to make such communications over as large a distance as we can conceive on the Earth, so that the population of the planet is more closely connected that at any time in our collective history. During your day, you might also travel on public transport or buy various things, each time tapping your bank card on devices that can read it. You will be bombarded every second by images, sounds, smells, tastes and other physical sensations, all of which are interpreted by your brain in highly complex ways. You will, as a living organism, have a subtle and ever-changing relationship with your immediate environment, fuelled by constant chemical reactions that proceed without your having to know about them, but which even give rise to your emotions. As the day ends, you might think about your various experiences during the day, possibly in a warm place conducive to sleep. If you instead look at the night sky, you will see some very interesting little twinkling dots, and if you travelled closer to them, you would see great plumes of swirling material that erupt out of the surface of these objects, before looping back in. Occasionally our own Sun does this, and if you were to travel close to the North pole, you would see beautiful coloured lights in the sky the following night.

All of the things described above — every single one of them — involve (at least in principle) the five very small equations we wrote above. This encapsulation of so much by apparently so little is a hallmark of modern physics, and was a culmination of literally thousands of years of effort by the human race, from all corners of the globe. Whatever culture you yourself most strongly identify with, it will certainly have influenced the path towards the remarkable equations stated above. What is also true is that the programme of unification of phenomena (i.e. electricity and magnetism) inherent in the Maxwell equations has continued to this day: the weak and strong nuclear forces were unified with electromagnetism into a single consistent theory (the Standard Model of Particle Physics) as recently as the 1970s. The experimental verification of this theory was arguably completed with the discovery of the Higgs boson in 2012. However, we have reason to believe that further unification is possible — and indeed necessary. Gravity (as described by General Relativity) is not yet unified with the other forces, and the inability to do so hampers our ability to understand the very early universe, or the centre of black holes.

The Maxwell equations also directly anticipated the two great revolutions of 20th century physics — relativity and quantum mechanics. The equations turn out to be expressible in a form which is manifestly invariant under the well-known Lorentz transformations of Special Relativity (discussed in the following chapter). Thus, they "know about" Special Relativity, even though this had not yet been discovered. Secondly, the equations have an abstract mathematical symmetry (discussed here in Chapter 9), that makes a consistent quantum theory of the electromagnetic field possible. The techniques used to quantise the theory also apply to the other forces, and matter particles, in nature.

Returning to our study of the Maxwell equations, they are written above in the so-called *integral form*, in which explicit line and surface integrals are featured. This obscures one of the properties that the use of fields was meant to circumvent: namely, in order to avoid the idea of action at a distance, it must be possible to write a field theory purely in terms of *local* quantities, defined around a single spacetime point. To this end, we can rewrite the Maxwell equations in so-called *differential form*, which makes locality manifest. However, in order to do this, we first have to learn some more vector calculus.

## 7.3 The Divergence

In Section 3.6, we saw how the flux of a vector field $\boldsymbol{F}(\boldsymbol{x})$ through some closed surface $S$ tells us how much abstract "stuff" associated with the field is flowing out of the volume $V$ enclosed by $S$. If the flux is non-zero, this stuff must be somehow *created* in the volume $V$, i.e. there must be a *source* for the field. This idea is indeed encapsulated by Gauss' law: the flux of the electric field is non-zero only if there is charge inside the surface, and it is thus charge that creates the electric field. Likewise, the flux of the magnetic field through any closed surface is zero, as there are no isolated sources of magnetic fields.

Now consider taking a closed surface $S$ around a point enclosing volume $V$, but shrinking the surface smaller and smaller, so that it becomes infinitesimally small. The volume $V$ likewise becomes small, and it turns out that the flux of a vector field through $S$, normalised by the volume $V$, converges to a finite number. This is known as the *divergence* of the vector field $\boldsymbol{F}$, and may be defined formally as

$$\lim_{V\to 0}\frac{1}{V}\oiint d\boldsymbol{S}\cdot\boldsymbol{F}, \tag{7.10}$$

where the symbol "$\lim_{V\to 0}$" denotes taking the volume (and surface $S$) smoothly to zero around a given point $\boldsymbol{x}$. The divergence is a *scalar*, that measures how much stuff is flowing away from a given point in space. Examples are shown in Figure 7.2. In Figure 7.2(a) of the figure, the field is diverging away from the point, implying a non-zero positive divergence. In Figure 7.2(b), the field is flowing towards the point, leading to a non-zero, but *negative* divergence (i.e. the point represents a *sink* of the field, rather than a source). Finally, Figure 7.2(c) depicts a field in which there is no net flow of stuff towards or away from the given point, thus a zero divergence. We will see in more detail below why the divergence is useful for our purposes (i.e. that of rewriting Maxwell's equations). But first let us find an explicit expression for the divergence in terms of the components of the vector field, as follows. Given that we are taking the surface $S$ to

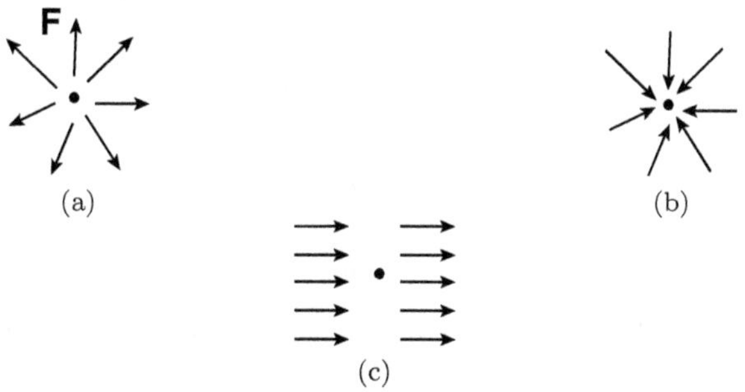


Fig. 7.2 (a) A point at which the divergence of the vector field $\boldsymbol{F}$ is positive; (b) a point at which the divergence is negative; (c) a point at which the divergence is zero.

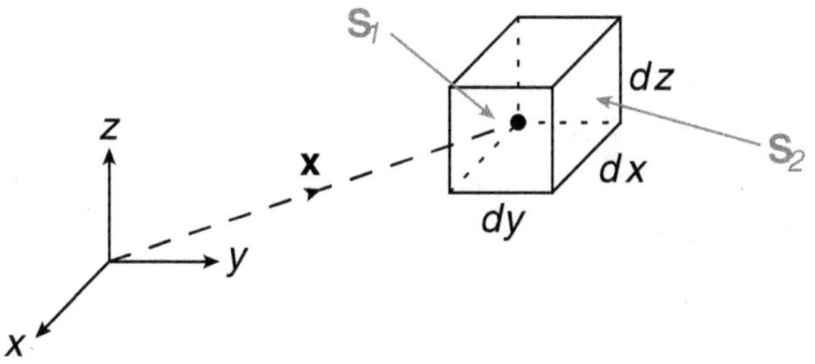


Fig. 7.3 A rectangular prism around the point $\boldsymbol{x}$.

zero, we can define any shape for this surface. Let us, for convenience, take a small rectangular prism as shown in Figure 7.3. Let the point

$$\boldsymbol{x} = \begin{pmatrix} x \\ y \\ z \end{pmatrix}$$

be situated at the back-left of the prism, as shown. Provided the volume

$$dV = dx\, dy\, dz$$

enclosed by the prism is sufficiently small, we can assume that the vector field $\boldsymbol{F}$ is approximately constant on each face of the prism. Let us examine the flux on the face $S_1$ labelled in the figure. The area of this face is $dydz$, and a normal vector pointing out of the prism points in the negative $x$

direction, so the flux is

$$\begin{aligned}\boldsymbol{F}\cdot d\boldsymbol{S}_1 &= \boldsymbol{F}\cdot(-\hat{\boldsymbol{\imath}})dydz\\ &= -\frac{F_x(x,y,z)}{dx}dxdydz\\ &= -\frac{F_x(x,y,z)}{dx}dV,\end{aligned}$$

where $F_x$ denotes the $x$-component of $\boldsymbol{F}$. Similarly, the flux through the surface $S_2$ is

$$\begin{aligned}\boldsymbol{F}\cdot d\boldsymbol{S}_2 &= \boldsymbol{F}\cdot\hat{\boldsymbol{\imath}}\,dydz\\ &= \frac{F_x(x+dx,y,z)}{dx}dV,\end{aligned}$$

where we must be careful in noting that the field on the face $S_2$ is defined at $x+dx$, rather than $x$. We do not have to worry about the variation of $y$ and $z$ on the face $S_2$, as we are taking the field to be constant on the face: if it isn't, we can simply take $dy$ and $dz$ to be smaller. Adding together the above two contributions, the total flux through faces $S_1$ and $S_2$ is

$$\left(\frac{F_x(x+dx,y,z)-F_x(x,y,z)}{dx}\right)dV \quad \xrightarrow{dx\to 0} \quad \frac{\partial F_x}{\partial x}dV,$$

where we have recognised the definition of the partial derivative of $F_x$ with respect to $x$ as $dx$ is taken infinitesimally small. Similarly, one may show that the flux through the upper/lower surfaces is

$$\frac{\partial F_z}{\partial z}dV,$$

and through the front/back surfaces

$$\frac{\partial F_y}{dy}dV.$$

Adding together all contributions, the total flux through the infinitesimal prism is

$$\oiint_S d\boldsymbol{S}\cdot\boldsymbol{F} = \left(\frac{\partial F_x}{\partial x}+\frac{\partial F_y}{\partial y}+\frac{\partial F_z}{\partial z}\right)dV.$$

From the above definition, the divergence is then

$$\lim_{dV\to 0}\frac{1}{dV}\oiint_S d\boldsymbol{S}\cdot\boldsymbol{F} = \frac{\partial F_x}{\partial x}+\frac{\partial F_y}{\partial y}+\frac{\partial F_z}{\partial z}.$$

This is usually written as $\nabla \cdot \boldsymbol{F}$, where $\nabla$ was introduced in Eq. (3.34). Indeed, following the usual notation for the dot product implies

$$\nabla \cdot \boldsymbol{F} = \begin{pmatrix} \partial/\partial x \\ \partial/\partial y \\ \partial/\partial z \end{pmatrix} \cdot \begin{pmatrix} F_x \\ F_y \\ F_z \end{pmatrix} = \frac{\partial F_x}{\partial x} + \frac{\partial F_y}{\partial y} + \frac{\partial F_z}{\partial z} \tag{7.11}$$

as required. Sometimes, Eq. (7.11) is quoted as the *definition* of the divergence. However, this is not the definition! Were I to simply give you Eq. (7.11), I feel confident in saying that you would have no idea at all what this was meant to represent. Instead, Eq. (7.11) is merely the expression for the divergence if we are in Cartesian coordinates. Different formulae for the divergence can be obtained in other coordinate systems (e.g. spherical or cylindrical polar coordinates), but are beyond the scope of this book. The actual definition of the divergence is, of course, Eq. (7.10), which explicitly tells us what it is: a measure of the "amount" of vector field leaving a given point in space.

In general, the divergence depends on position, i.e. $\nabla \cdot \boldsymbol{F} = f(\boldsymbol{x})$ for some *scalar function* $f(\boldsymbol{x})$. Although the divergence is a property of infinitesimal surfaces, we can derive a useful formula involving the divergence, that applies for finite surfaces. Consider a large surface $S$, enclosing a volume $V$. We can divide this volume into (infinitely many) infinitesimal cubes, that completely fill the volume. By the definition of the divergence, the flux through each infinitesimal cube is

$$\nabla \cdot \boldsymbol{F} dV,$$

and we can consider summing over all such cubes, i.e. constructing the volume integral

$$\int\int\int_V dV\, \nabla \cdot \boldsymbol{F}.$$

If we take a cube somewhere in the centre of the volume, it will share faces with neighbouring cubes, an example of which is shown in Figure 7.4. On the shared face shown, the vector area is pointing to the right for the left-hand cube, but to the left for the right-hand cube. Given that the field is constant on the face, it follows that the contribution to the total flux (summed over cubes) *cancels* on this face, when we add both cubes together. This same argument applies to *any* shared face, so that upon summing over all cubes, the only flux that contributes is that due to the

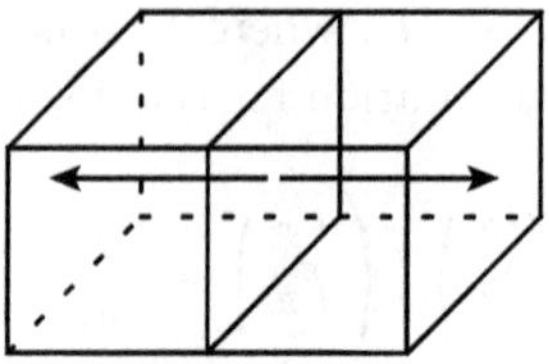

Fig. 7.4 Two cubes sharing a common face.

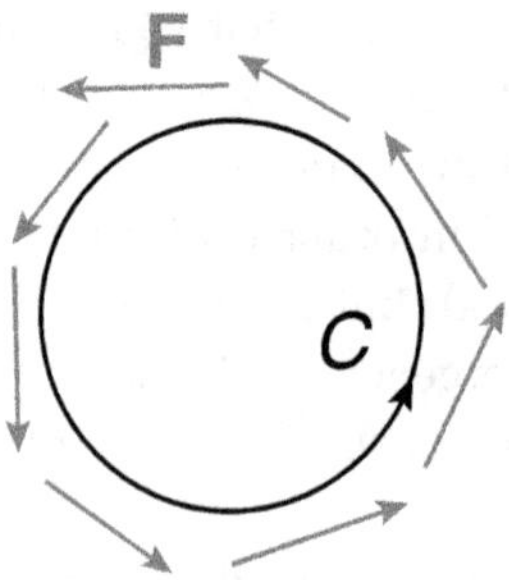


Fig. 7.5 A vector field rotating around a closed contour $C$.

outer edges of the cubes that define the boundary surface $S$. That is, we must have

$$\int\int\int dV\, \nabla \cdot \boldsymbol{F} = \oiint_S d\boldsymbol{S} \cdot \boldsymbol{F}. \tag{7.12}$$

In words: the volume integral of the divergence of a vector field $\boldsymbol{F}$ throughout a volume $V$ is equal to the surface integral of $\boldsymbol{F}$ over a closed surface $S$ that encloses $V$. This is called the *divergence theorem*, and is a very useful result: it relates the flux through a surface (a *global* quantity) to the divergence (a *local quantity*). We will use this theorem when rewriting Gauss' law in differential form.

## 7.4 The Curl

Consider the line integral around a closed curve $C$, as exemplified by Figure 7.5. Clearly the line integral will be nonzero only if there is a net *rotational* character of the field around the loop, and indeed the figure represents an extreme example in which the field everywhere has a component in the same anticlockwise direction. The rotation itself has a vector nature, in that it only makes sense to talk about the amount of rotation around a particular axis, which in this case points out of the page.

We may now imagine taking the loop smaller and smaller around a given point $\boldsymbol{x}$. If we do so, it turns out that the line integral per unit area bounded by the loop converges to a finite number. This number measures how much the field $\boldsymbol{F}$ rotates about the point $\boldsymbol{x}$, and we can make this fully general as follows. Consider a unit vector $\hat{\boldsymbol{n}}$, which denotes the direction of a hypothetical rotation axis. Then we can define a vector $\mathbf{c}$ called *the curl*,[1] and such that

$$\mathbf{c}\cdot\hat{\boldsymbol{n}} = \lim_{A\to 0}\frac{1}{A}\oint_C \boldsymbol{F}\cdot d\boldsymbol{l}.$$

On the right-hand side is a line integral of $\boldsymbol{F}$ around a closed loop bounding an area $A$, oriented perpendicular to the axis $\hat{\boldsymbol{n}}$. Thus, the dot product on the left measures the rotational aspect of the field around the axis $\hat{\boldsymbol{n}}$ at the point $\boldsymbol{x}$. The fact that $\mathbf{c}$ is a vector means that we can measure this rotation for *any* direction $\hat{\boldsymbol{n}}$. As an example, let us take $\hat{\boldsymbol{n}} = \hat{\boldsymbol{k}}$ (i.e. a unit vector in the $z$ direction), and an infinitesimal loop in the $(x, y)$ plane, as shown in Figure 7.6. The point $\boldsymbol{x}$ is at the bottom-left corner of the loop, and if the loop is small enough then we can take the field $\boldsymbol{F}$ to be constant on each

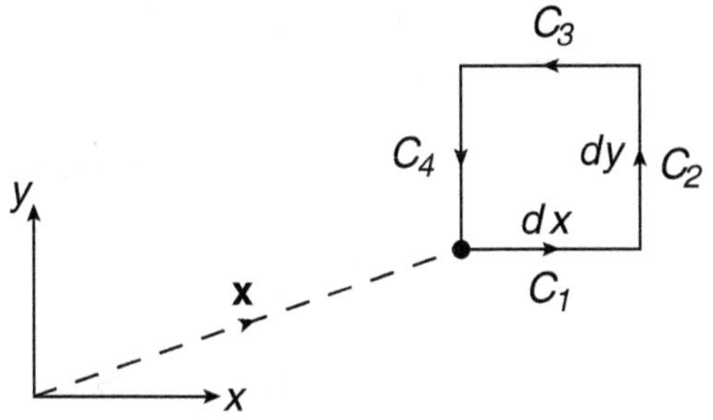


Fig. 7.6 An infinitesimal loop in the $(x, y)$ plane.

[1]The vector $\mathbf{c}$ is my own notation for the curl, and will not be found in any other textbooks. The reason for this is that we will use a better notation shortly, but we need to derive it first!

segment. The line integral of $\boldsymbol{F}$ along the segment $C_1$ is then given by

$$\int_{C_1} \boldsymbol{F}\cdot d\boldsymbol{l} = F_x(x,y,z)dx$$
$$= \frac{F_x(x,y,z)}{dy}dxdy$$
$$= \frac{F_x(x,y,z)}{dy}dA,$$

where $dA = dxdy$ is the area of the loop. Similarly, the line integrals on the other segments give

$$\int_{C_2} \boldsymbol{F}\cdot d\boldsymbol{l} = \frac{F_y(x+dx,y,z)}{dx}dA;$$
$$\int_{C_3} \boldsymbol{F}\cdot d\boldsymbol{l} = -\frac{F_x(x,y+dy,z)}{dy}dA;$$
$$\int_{C_4} \boldsymbol{F}\cdot d\boldsymbol{l} = -\frac{F_y(x,y,z)}{dx}dA,$$

so that the complete line integral around the loop is

$$\oint_C \boldsymbol{F}\cdot d\boldsymbol{l} = \left[\left(\frac{F_y(x+dx,y,z)-F_y(x,y,z)}{dx}\right) - \left(\frac{F_x(x,y+dy,z)-F_x(x,y,z)}{dy}\right)\right]dA$$
$$\xrightarrow{dx,dy\to 0} \left(\frac{\partial F_y}{\partial x} - \frac{\partial F_x}{\partial y}\right)dA.$$

From the above definition, it follows that the $z$-component of the curl is

$$c_z = \frac{1}{dA}\oint_C \boldsymbol{F}\cdot\boldsymbol{l} = \frac{\partial F_y}{\partial x} - \frac{\partial F_x}{\partial y}.$$

A similar analysis in the $(x,z)$ and $(y,z)$ planes gives

$$c_x = \frac{\partial F_y}{\partial z} - \frac{\partial F_z}{\partial y}, \quad c_y = \frac{\partial F_x}{\partial z} - \frac{\partial F_z}{\partial x}.$$

This is usually written as

$$\boldsymbol{c} \equiv \nabla\times\boldsymbol{F},$$

which indeed reproduces the right formula. That is, the usual rule for the cross-product gives

$$\nabla \times \boldsymbol{F} = \begin{pmatrix} \partial/\partial x \\ \partial/\partial y \\ \partial/\partial z \end{pmatrix} \times \begin{pmatrix} F_x \\ F_y \\ F_z \end{pmatrix} = \begin{pmatrix} \partial F_z/\partial y - \partial F_y/\partial z \\ \partial F_x/\partial z - \partial F_z/\partial x \\ \partial F_y/\partial x - \partial F_x/\partial y \end{pmatrix} \tag{7.13}$$

as required. The curl is perhaps more complicated to think about than the divergence. However, there is a nice piece of physical intuition that you can apply to determine whether or not the curl should be zero (in some direction) for a given vector field. Consider taking the components of a vector field in a plane perpendicular to a given direction (which we will take out of the page). Then imagine placing a small cross in the vector field, as shown in Figure 7.7, and also that the arrows represent the flow of a fluid, which can hit the cross and make it move. If the cross *rotates*, this means that the component of the curl out of the page is nonzero. It will be positive (negative), if the cross rotates anticlockwise (clockwise). If there is no rotation, as in Figure 7.7(b), then the component of the curl out of the page is zero. Another thing that can happen, of course, is that the cross is carried along with the fluid, in such a way that it does not rotate. Then, also, the curl is zero.

The definition of the curl involves a line integral around an infinitesimal closed loop. We can, however, derive a useful formula involving the curl for large loops. Any given macroscopic contour $C$ bounds some area, which can be partitioned into (infinitely many) infinitesimal loops, as shown in Figure 7.8. Let us take each infinitesimal loop to have vector area $d\boldsymbol{S}$. By the definition of the curl, the line integral of some vector field $\boldsymbol{F}$ around each loop is

$$d\boldsymbol{S} \cdot \nabla \times \boldsymbol{F},$$

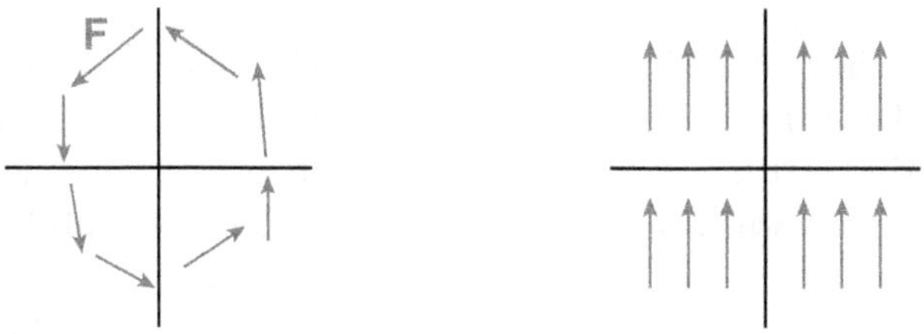


(a) (b)

Fig. 7.7 A cross placed into a vector field: (a) a point where the curl component out of the page is nonzero; (b) a point at which the curl component out of the page is zero.

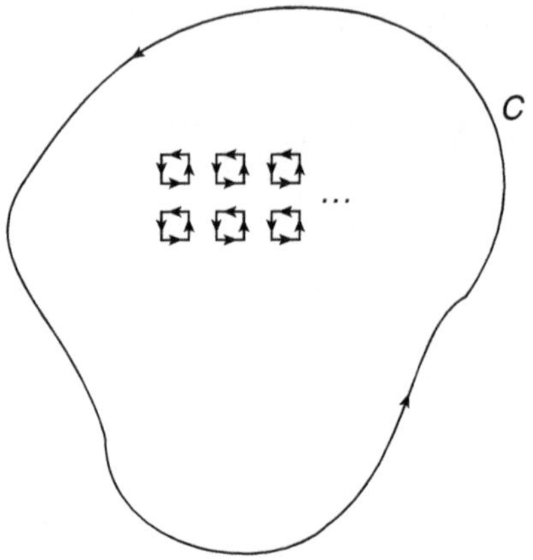


Fig. 7.8 One may partition the area bounded by a closed contour $C$ into infinitesimal loops.

and we can then see what happens when we sum over all such loops. Firstly, this gives (by definition) the surface integral

$$\int\int d\boldsymbol{S}\cdot\nabla\times\boldsymbol{F}.$$

However, in summing over the individual line integrals, the integral over any edge which is shared by two loops *cancels*, as $d\boldsymbol{l}$ on such a segment points in opposite directions in the two loops, but the field is the same. This leaves only the sum of line integrals on segments which are not shared, namely over the segments comprising the outer contour $C$, and we are led to the result

$$\int\int_S d\boldsymbol{S}\cdot\nabla\times\boldsymbol{F}=\oint_C \boldsymbol{F}\cdot d\boldsymbol{l}. \tag{7.14}$$

In words: the line integral of a vector field $\boldsymbol{F}(\boldsymbol{x})$ around a closed contour $C$ is equal to the surface integral of the curl of $\boldsymbol{F}(\boldsymbol{x})$ over a surface bounded by $C$. Scrutiny of the above argument indeed reveals that the surface $S$ can be *any* surface which spans $C$, and another way to see this is as follows. Consider two surfaces $S_1$ and $S_2$ that both span a given contour $C$, as in Figure 5.15. From our analogy of flux as representing a "flow of stuff" across a surface, all of the stuff flowing through $S_1$ will be captured by $S_2$, provided no stuff is created or destroyed in between. Thus, there must be no points in between $S_1$ and $S_2$ where there is a net amount of stuff flowing in or out. In other words, the flux of $\nabla \times \boldsymbol{F}$ through $S_1$ and $S_2$ will be the same, provided the divergence of $\nabla \times \boldsymbol{F}$ vanishes everywhere in between the two surfaces. That is,

$$\int\int_{S_1} d\boldsymbol{S}_1 \cdot (\nabla \times \boldsymbol{F}) = \int\int_{S_2} d\boldsymbol{S}_2 \cdot (\nabla \times \boldsymbol{F}),$$

provided

$$\nabla \cdot (\nabla \times \boldsymbol{F}) = 0. \tag{7.15}$$

From the definitions of the divergence and the curl, you can indeed prove that Eq. (7.15) holds in full generality, for any physically well-behaved vector field $\boldsymbol{F}$ (i.e. one whose partial derivatives exist).

## 7.5 Maxwell Equations in Differential Form

Having taken a detour to learn about the divergence and the curl, we can now return to finding a local form of Maxwell's equations. The first Maxwell equation is Eq. (7.5), where $Q$ represents the total charge enclosed in the volume $V$ bounded by the closed surface $S$. If we know the *charge density* (or charge per unit volume) $\rho(\boldsymbol{x})$, we can write the total charge as the volume integral

$$Q = \int\int\int_V \rho(\boldsymbol{x}) dV,$$

so that Eq. (7.5) becomes

$$\oiint_S \boldsymbol{E} \cdot d\boldsymbol{S} = \frac{1}{\epsilon_0} \int\int\int_V \rho(\boldsymbol{x}) dV.$$

We may compare this with the divergence theorem, Eq. (7.12) which, like Eq. (7.5), must be true for any surface $S$ and volume $V$. Then the only way that the two equations can be consistent is if

$$\nabla \cdot \boldsymbol{E} = \frac{\rho(\boldsymbol{x})}{\epsilon_0}.$$

This tells us that the divergence of the electric field at a point $\boldsymbol{x}$ depends on the charge density at the same point. It is thus a completely local equation, depending only on properties near a single point. Likewise, the second

Maxwell equation — Eq. (7.6) implies

$$\nabla \cdot \boldsymbol{B} = 0.$$

The third Maxwell equation is Eq. (7.7), and we may rewrite this as

$$\oint_C \boldsymbol{E} \cdot d\boldsymbol{l} = -\frac{\partial}{\partial t} \int\int_S d\boldsymbol{S} \cdot \boldsymbol{B},$$

to make clear that the flux is defined in terms of a surface integral.[2] We may compare this equation with Stokes' theorem, where both equations are meant to hold for arbitrary contours $C$, bounding arbitrary surfaces $S$. The only way that this can be consistent is if

$$\nabla \times \boldsymbol{E} = -\frac{\partial \boldsymbol{B}}{\partial t}.$$

Finally, the fourth Maxwell equation is Eq. (7.8). The first term on the right-hand side involves the total current flowing through a surface $S$ bounded by the closed contour $C$ on the left-hand side, and we may write this as an explicit surface integral by remembering that the *current density* $\boldsymbol{J}$ is defined to be a vector such that $|\boldsymbol{J}|$ is the current flowing per unit area perpendicular to $\boldsymbol{J}$. Thus, we can write the total current flowing through a surface $S$ as

$$I = \int\int_S d\boldsymbol{S} \cdot \boldsymbol{J}.$$

Also, the rate of change of flux in the displacement current (second term on the right-hand side of Eq. (7.8)) can be written as

$$\frac{d\Phi_E}{dt} = \frac{\partial}{\partial t} \int\int_S d\boldsymbol{S} \cdot \boldsymbol{E},$$

so that Eq. (7.8) assumes the form

$$\oint_C \boldsymbol{B} \cdot d\boldsymbol{l} = \int\int_S d\boldsymbol{S} \cdot \left[\mu_0 \boldsymbol{J} + \mu_0 \epsilon_0 \frac{\partial \boldsymbol{E}}{\partial t}\right].$$

Comparing this with Stokes' theorem implies

$$\nabla \times \boldsymbol{B} = \mu_0 \boldsymbol{J} + \mu_0 \epsilon_0 \frac{\partial \boldsymbol{E}}{\partial t}.$$

[2]You may worry about the fact that I have replaced a total derivative with respect to time in Eq. (7.7) with a partial derivative acting on the surface integral. This is indeed the correct thing to do, given that the surface defining the flux is held stationary as the derivative with respect to time is taken.

Again this is a local equation — it relates the curl of the magnetic field at a point to the electric field and the flow of current at the same point. In summary, we have found the four local equations

$$\nabla \cdot \boldsymbol{E} = \frac{\rho}{\epsilon_0}, \tag{7.16}$$

$$\nabla \cdot \boldsymbol{B} = 0, \tag{7.17}$$

$$\nabla \times \boldsymbol{E} = -\frac{\partial \boldsymbol{B}}{\partial t}, \tag{7.18}$$

$$\nabla \times \boldsymbol{B} = \mu_0 \boldsymbol{J} + \mu_0 \epsilon_0 \frac{\partial \boldsymbol{E}}{\partial t}. \tag{7.19}$$

These are *Maxwell's equations in differential form.* That is, expanding out the definitions of the divergence and curl, one finds a set of partial differential equations for the fields, rather than integral equations. It is worth pointing out that although the form of the divergence and the curl may be different in different coordinate systems, the form of Eqs. (7.16)–(7.19) is valid in *any* coordinate system.

We may also ask whether the integral and differential Maxwell equations are *always* equivalent to each other. The answer to this is actually no: in deriving the differential form, we have assumed that we can use the divergence and Stokes theorems. These theorems are well-behaved in simple spaces (such as the 3D Euclidean space in which we seem to live), but would need modification in more complicated situations, e.g. spaces with holes cut out. The latter are not as esoteric as they sound — we might model certain materials, for example, by saying that the properties of space have been modified. For this reason, we usually take the differential form of Maxwell's equations to be the fundamental definition of electromagnetism, as they depend only on local properties. Whether or not they scale up to the integral form then depends on the *global* properties of the spacetime.

You may be wondering how we can be confident that the above equations are all we need to describe electromagnetism. The fact we have rewritten them in differential form allows us to see this quite nicely. It turns out (in three space dimensions!) that once we have specified the divergence of a vector field (i.e. how it spreads out at a point) and its curl (how it rotates around a point), its behaviour is *completely characterised.* If you doubt this, go to the park and watch some dogs. Each dog has a velocity vector, and thus each dog gives you an arrow in a "vector field" of dogs! You will see them running away from or towards each other (a non-zero divergence of dogs), or chasing their own tails (a non-zero curl of dogs). But you won't see any behaviour that corresponds to something *not* captured by the divergence or the curl! Of course, this example is really 2D, given that dogs can't fly. But you can look at flocks of birds to be convinced that the divergence and curl are all you need in three space dimensions.

Given the above observation, we can see why the four Maxwell equations are all we need. To fully describe electromagnetism, we need a vector field for both electricity ($\boldsymbol{E}$) and magnetism ($\boldsymbol{B}$). Then we need to specify the divergence and curl of each, which is exactly what Eqs. (7.16)–(7.19) do!

## 7.6 Electromagnetic Waves

Having finally arrived at Maxwell's equations, let us examine one of their most profound consequences. First, let us note that a complicated, but nevertheless true, equation is

$$\nabla \times (\nabla \times \boldsymbol{F}) = \nabla(\nabla \cdot \boldsymbol{F}) - \nabla^2 \boldsymbol{F}, \tag{7.20}$$

where in the last term on the right-hand side we have introduced the notation

$$\nabla^2 \boldsymbol{F} = \frac{\partial^2 \boldsymbol{F}}{\partial x^2} + \frac{\partial^2 \boldsymbol{F}}{\partial y^2} + \frac{\partial^2 \boldsymbol{F}}{\partial z^2}, \tag{7.21}$$

as suggested by the notation for $\nabla$ itself.[3] One way to prove Eq. (7.20) is simply by brute force. For example, the definitions of the curl and divergence imply that the $x$-component of the left-hand side is

$$\begin{aligned}
[\nabla \times (\nabla \times \boldsymbol{F})]_x &= \frac{\partial}{\partial y}(\nabla \times \boldsymbol{F})_z - \frac{\partial}{\partial z}(\nabla \times \boldsymbol{F})_y \\
&= \frac{\partial}{\partial y}\left(\frac{\partial F_y}{\partial x} - \frac{\partial F_x}{\partial y}\right) - \frac{\partial}{\partial z}\left(\frac{\partial F_x}{\partial z} - \frac{\partial F_z}{\partial x}\right) \\
&= \frac{\partial^2 F_y}{\partial x \partial y} - \frac{\partial^2 F_x}{\partial y^2} - \frac{\partial^2 F_x}{\partial z^2} + \frac{\partial^2 F_z}{\partial z \partial x}.
\end{aligned}$$

[3]The operator $\nabla^2 = \frac{\partial^2}{\partial x^2} + \frac{\partial^2}{\partial y^2} + \frac{\partial^2}{\partial z^2}$ is commonly called the *Laplacian operator*, and can be applied to scalar functions as well as vectors.

On the other hand, the $x$-component of the right-hand side of Eq. (7.20) is

$$\frac{\partial}{\partial x}\left(\frac{\partial F_x}{\partial x}+\frac{\partial F_y}{\partial y}+\frac{\partial F_z}{\partial z}\right)-\frac{\partial^2 F_x}{\partial x^2}-\frac{\partial^2 F_x}{\partial y^2}-\frac{\partial^2 F_x}{\partial z^2}$$
$$=\frac{\partial^2 F_y}{\partial x\partial y}-\frac{\partial^2 F_x}{\partial y^2}-\frac{\partial^2 F_x}{\partial z^2}+\frac{\partial^2 F_z}{\partial z\partial x},$$

in agreement with the above result. A similar analysis can be used to show that the other components work too.

Now consider Maxwell's equations in a vacuum, i.e. where $\rho = \boldsymbol{J} = 0$ at all points throughout space:

$$\nabla \cdot \boldsymbol{E} = 0, \tag{7.22}$$
$$\nabla \cdot \boldsymbol{B} = 0, \tag{7.23}$$
$$\nabla \times \boldsymbol{E} = -\frac{\partial \boldsymbol{B}}{\partial t}, \tag{7.24}$$
$$\nabla \times \boldsymbol{B} = \mu_0\epsilon_0\frac{\partial \boldsymbol{E}}{\partial t}. \tag{7.25}$$

If we take the curl of Eq. (7.24), we get (using Eq. (7.20))

$$\begin{aligned}\nabla \times (\nabla \times \boldsymbol{E}) &= \nabla(\nabla \cdot \boldsymbol{E}) - \nabla^2\boldsymbol{E}\\ &= -\frac{\partial}{\partial t}(\nabla \times \boldsymbol{B})\\ &= -\mu_0\epsilon_0\frac{\partial^2 \boldsymbol{E}}{\partial t^2},\end{aligned}$$

where we have used Eq. (7.25) in the last line. We can use Eq. (7.22) to get rid of the first term on the right-hand side of the first line, leaving

$$-\nabla^2\boldsymbol{E} = -\mu_0\epsilon_0\frac{\partial^2 \boldsymbol{E}}{\partial t^2}.$$

Similarly, we can take the curl of Eq. (7.25), and we get

$$\begin{aligned}\nabla \times (\nabla \times \boldsymbol{B}) &= \nabla(\nabla \cdot \boldsymbol{B}) - \nabla^2\boldsymbol{B}\\ &= \mu_0\epsilon_0\frac{\partial}{\partial t}(\nabla \times \boldsymbol{E}) = -\mu_0\epsilon_0\frac{\partial^2 \boldsymbol{B}}{\partial t^2}.\end{aligned}$$

Using Eq. (7.23) then implies that both $\boldsymbol{E}$ and $\boldsymbol{B}$ obey a similar equation:

$$\nabla^2\boldsymbol{E} = \frac{1}{c^2}\frac{\partial^2 \boldsymbol{E}}{\partial t^2}, \quad \nabla^2\boldsymbol{B} = \frac{1}{c^2}\frac{\partial^2 \boldsymbol{B}}{\partial t^2}, \tag{7.26}$$

where

$$c = \frac{1}{\sqrt{\mu_0 \epsilon_0}}. \tag{7.27}$$

To look for a solution, let us make the ansatz

$$\boldsymbol{E} = E_y(x,t)\,\hat{\boldsymbol{j}}$$

(i.e. the electric field depends only on $x$ and $t$, and points in the $y$-direction). Equation (7.26) then reduces to

$$\frac{\partial^2 E_y(x,t)}{\partial t^2} = c^2 \frac{\partial^2 E_y(x,t)}{\partial x^2}. \tag{7.28}$$

A solution of this equation is

$$E_y = E_0 \cos(kx - \omega t), \quad \omega = ck, \tag{7.29}$$

for some constants $E_0$ and $k$. To check this, note that differentiating $E_y$ once with respect to $t$ gives

$$\frac{\partial E_y}{\partial t} = \omega E_0 \sin(kx - \omega t),$$

so that differentiating a second time gives

$$\frac{\partial^2 E_y}{\partial t^2} = -\omega^2 E_0 \cos(kx - \omega t). \tag{7.30}$$

Furthermore, we have

$$\frac{\partial E_y}{\partial x} = -kE_0 \sin(kx - \omega t)$$

so that

$$\frac{\partial^2 E_y}{\partial x^2} = -k^2 E_0 \cos(kx - \omega t).$$

This is indeed equal to Eq. (7.30) provided $\omega = ck$.

Given the solution for $\boldsymbol{E}$, we can find $\boldsymbol{B}$ using Eq. (7.24):

$$\frac{\partial \boldsymbol{B}}{\partial t} = -\nabla \times \boldsymbol{E} = -\begin{pmatrix} \partial E_z/\partial y - \partial E_y/\partial z \\ \partial E_x/\partial z - \partial E_z/\partial x \\ \partial E_y/\partial x - \partial E_x/\partial y \end{pmatrix} = \begin{pmatrix} 0 \\ 0 \\ -\partial E_y/\partial x \end{pmatrix}.$$

We therefore find that the magnetic field must point in the $z$-direction:

$$\boldsymbol{B} = B_z \hat{\boldsymbol{k}},$$

where $B_z$ must satisfy the equation

$$\frac{\partial B_z}{\partial t} = -\frac{\partial E_y}{\partial x},$$

implying

$$B_z = -\int dt \frac{\partial E_y}{\partial x} = \int dt k E_0 \sin(kx - \omega t) = \frac{E_0}{c} \cos(kx - \omega t),$$

where we have ignored an arbitrary integration constant, and used the definition of $c$ from Eq. (7.29). Putting things together, we have found the specific solution

$$\boldsymbol{E} = E_0 \cos(\omega t - kx)\, \hat{\boldsymbol{j}}, \quad \boldsymbol{B} = \frac{E_0}{c} \cos(\omega t - kx)\, \hat{\boldsymbol{k}}. \tag{7.31}$$

The interpretation of this result is relatively straightforward: it is a wave moving in the $x$-direction! That is, both the electric and magnetic fields oscillate with position and time, and the crests of the wave move with a speed $c$. The electric and magnetic fields are *transverse* to the direction of the wave, and wavefronts (surfaces of constant phase) are planes perpendicular to the direction of travel. Thus, our example is a so-called *plane wave*.

Note that, although $\omega$ and $k$ are not independent, there is no restriction on what value of $\omega$ we take, provided we do not also fix $k$. Thus, we have actually found a whole family of solutions, each of which has a unique frequency

$$f = \frac{\omega}{2\pi}. \tag{7.32}$$

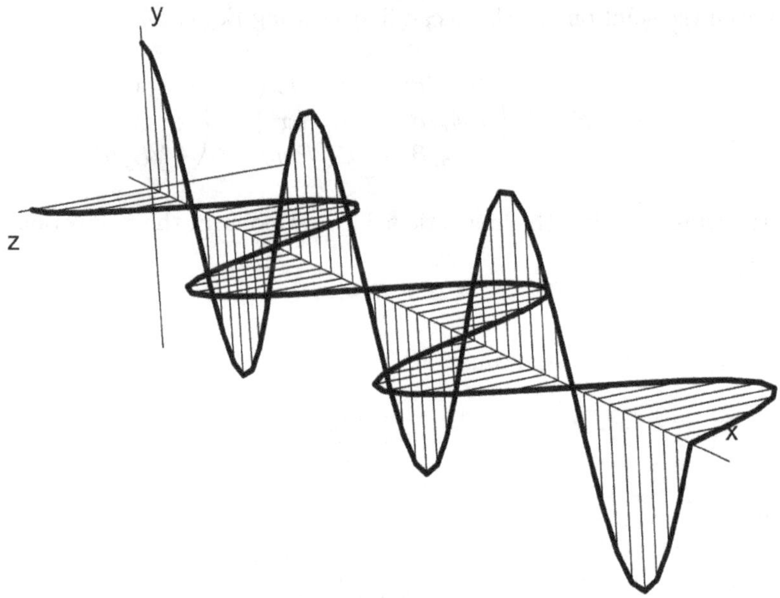


Fig. 7.9 The electric and magnetic fields (in the $y$ and $z$ directions, respectively) for a particular electromagnetic wave moving in the positive $x$-direction.

Equivalently, if we do not specify $\omega$, we have a choice of different wavelengths

$$\lambda = \frac{2\pi}{k}. \tag{7.33}$$

The speed of the waves is given in Eq. (7.27), and depends on the permittivity and permeability of free space. These are constants of nature, so that the speed of the waves is constant, and the same for all solutions. Although we have found a plane wave here, other wave-like solutions of Maxwell's equations are possible (e.g. waves with spherical wavefronts).

It is interesting to ponder what allows the wave-like solutions to exist. The basic reason is that Maxwell's equations imply that changing electric fields create magnetic fields, and vice versa. Thus, an oscillating electric field will set up an oscillating magnetic field, so that the oscillations propagate and become self-sustaining. That is, empty space can support the propagation of *electromagnetic radiation*. Depending on the wavelength $\lambda$, we call this radiation by different names. In order of increasing wavelength, the names include: $\gamma$ rays, X-rays, ultraviolet to infrared radiation (between which is the visible light spectrum), microwaves and radio waves.

Table 7.1 Different types of radiation lying on the electromagnetic spectrum, classified in order of increasing wavelength.

| Radiation type | Wavelength |
|---|---|
| $\gamma$ rays | <10 pm |
| X-rays | 10 pm–10 nm |
| Ultraviolet | 10 nm–400 nm |
| Visible light | 400 nm–750 nm |
| Infrared | 750 nm–1 mm |
| Microwaves | 1 mm–1 m |
| Radio waves | >1 m |

Approximate wavelengths defining these somewhat arbitrary distrinctions are collected in Table 7.1. There is a huge number of applications of the electromagnetic spectrum, not least the fact that we are able to detect the world in the first place!

Note that without the displacement current introduced by Maxwell (which leads to the non-zero right-hand side of Eq. (7.25)), the wave-like solutions would not exist: we need changing electric fields to source magnetic fields, as well as the converse. Thus, in unifying electricity and magnetism, Maxwell resolved a centuries old dispute regarding whether light was a particle or a wave. The development of physics since Maxwell has been similar, in that combining different ideas typically predicts something new. Unifying Special Relativity (SR) with quantum mechanics, for example, led to the prediction of antimatter. Unifying three of the fundamental forces into the Standard Model led to the prediction of the Higgs boson. Ironically, shortly after Maxwell explained the wave nature of light, it was indeed shown to have particle-like properties. However, this was due to quantum behaviour, and a quantum particle is not really the same as a classical particle — even in quantum mechanics, light can have wave-like properties. A full understanding needs quantum field theory, which we will talk about more in Chapter 9.

We saw above that Maxwell's equations predict a constant speed of light, which must be the same for all observers, if they experience the same laws of physics. This is a remarkable fact, as it goes against the laws of mechanics that were known at that time. It is this prediction of Maxwell's theory that led Einstein to develop the theory of SR, and it is in this sense that electromagnetism directly anticipated SR. One way to view SR, indeed, is that it is the consistent unification of mechanics with the laws of electromagnetism.

## 7.7 The Poynting Vector

We saw earlier that if electric and magnetic fields are present in a region of spacetime, there is an energy per unit volume

$$u = \frac{1}{2}\epsilon_0|\boldsymbol{E}|^2 + \frac{1}{2\mu_0}|\boldsymbol{B}|^2.$$

For the plane wave above, we saw that

$$|\boldsymbol{E}| = c|\boldsymbol{B}|,$$

where $c$ is given by Eq. (7.27). For the wave, we thus find an energy density

$$\begin{aligned} u &= \frac{1}{2}\left(\epsilon_0 c|\boldsymbol{E}||\boldsymbol{B}| + \frac{1}{\mu_0 c}|\boldsymbol{E}||\boldsymbol{B}|\right) \\ &= \frac{1}{2}\sqrt{\frac{\epsilon_0}{\mu_0}}|\boldsymbol{E}||\boldsymbol{B}| + \frac{1}{2}\sqrt{\frac{\epsilon_0}{\mu_0}}|\boldsymbol{E}||\boldsymbol{B}|. \end{aligned}$$

We see that the energy carried by the electric and magnetic fields is the same, at all times. The total energy density is

$$u = \sqrt{\frac{\epsilon_0}{\mu_0}}|\boldsymbol{E}||\boldsymbol{B}|.$$

Now consider an area $A$ perpendicular to the wave direction. In time $dt$, an amount $cdt$ of the wave will cross this area, tracing out a volume

$$dU = uAcdt,$$

which carries an energy

$$dU = udV = uAcdt.$$

The *intensity* $I$ of the wave is defined as the energy transmitted per unit time, per unit area (alternatively: the power per unit area). We thus have

$$I = \frac{1}{A}\frac{dU}{dt} = uc = \frac{1}{\mu_0}|\boldsymbol{E}||\boldsymbol{B}|.$$

From the explicit solution for the plane wave, Eq. (7.31), we can also note that

$$\boldsymbol{E} \times \boldsymbol{B} \propto \hat{\boldsymbol{\imath}}.$$

That is, the cross-product of the electric and magnetic fields is proportional to the direction of travel. We can thus define a vector quantity pointing in this direction, whose magnitude is the intensity of the wave:

$$\boldsymbol{S} = \frac{1}{\mu_0}\boldsymbol{E} \times \boldsymbol{B}. \tag{7.34}$$

This is called the *Poynting vector*, and turns out to be a very general result: for any solution of Maxwell's equations (including non-plane waves), the Poynting vector represents the direction of energy flow, where $|\boldsymbol{S}|$ is the associated intensity.

## 7.8 Polarisation

Our above solution for an electromagnetic wave is clearly very special. Not only is it a plane wave, but it has a particular direction of travel, and a particular direction for the electric field. Once the latter is fixed, the direction of the magnetic field is also fixed, given that this must be orthogonal to the electric field. More generally, however, there can be different directions for the electric field, or even an electric field vector that does not point in a particular fixed direction, but instead changes its direction with time. A given behaviour for the electric field vector is referred to as the *polarisation* of an electromagnetic wave.

To examine this concept more precisely, let us take an electromagnetic plane wave propagating in the $+z$-direction, where this is assumed to be out of the page. The $(x, y)$ plane then lies in the page, and constitutes the transverse plane with respect to the direction of travel. The (oscillating) electric field must lie in this plane. As it oscillates, it traces out a shape in the plane, and the case of Eq. (7.31) is shown in Figure 7.10. That is, this is a straight line in the $y$ direction, given the electric field is always pointing in the positive or negative $y$ direction. To describe the most general case, let us introduce unit vectors in the transverse plane:

$$\hat{e}_1 = \begin{pmatrix} 1 \\ 0 \\ 0 \end{pmatrix}, \quad \hat{e}_2 = \begin{pmatrix} 0 \\ 1 \\ 0 \end{pmatrix}. \tag{7.35}$$

In our particular case, these are equivalent to $\hat{\boldsymbol{\imath}}$ and $\hat{\boldsymbol{\jmath}}$, respectively (i.e. the unit vectors in the $x$ and $y$ directions). However, for an arbitrary electromagnetic wave in an arbitrary direction of travel, we can always construct a plane transverse the direction of the wave at a given point, and choose a

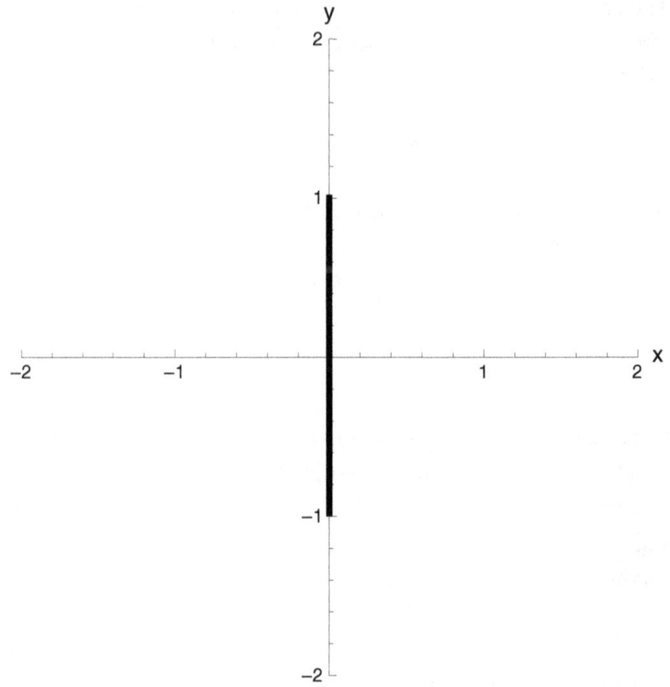


Fig. 7.10 The curve traced out in the plane transverse to the wave direction, by the electric field of Eq. (7.31).

coordinate system in this plane with basis vectors as in Eq. (7.35), hence our more general notation. Now note that the electric field of our wave solution of Eq. (7.31) can be written as

$$\boldsymbol{E} = \mathrm{Re}[E_0 e^{i(\omega t - kz)}]\hat{\boldsymbol{e}}_2. \tag{7.36}$$

Here, we have introduced a complex electric field in the square brackets, and taken the real part to yield the electric field of Eq. (7.31). We are always free to shift the origin of the time variable (or the spatial variables) if we want to. This amounts to adding a constant phase to the exponent of Eq. (7.36), such that this is also a solution. We are also free to rotate the coordinate system about the $z$-direction, thus redefining the $x$ and $y$ directions. It follows from this that an oscillation in *any* direction in the $(x, y)$ plane is also a solution. Finally, the fact that Maxwell's equations are linear implies that we can superpose two solutions to generate a third solution. Thus, the most general case of an electromagnetic plane wave will have an electric field of form

$$\boldsymbol{E} = \text{Re}\big[\big(A_1 e^{i\phi_1}\hat{\boldsymbol{e}}_1 + A_2 e^{i\phi_2}\hat{\boldsymbol{e}}_2\big)\, e^{i(\omega t - kz)}\big]. \tag{7.37}$$

That is, we may have a general amplitude $A_i$ in the direction of $\boldsymbol{e}_i$, plus a phase shift $\phi_i$ for the oscillation in that direction. If we want to, we can shift the origin of time in order to remove an overall phase (e.g. setting $\phi_1 = 0$). However, there would remain a potential phase difference between the first and second terms. We will choose to leave both phases arbitrary until further notice, and may then write Eq. (7.37) as

$$\boldsymbol{E} = \text{Re}\big[\boldsymbol{e}e^{i(\omega t - kz)}\big], \tag{7.38}$$

where

$$\boldsymbol{e} = \sum_{j=1}^{2} A_j e^{i\phi_j}\hat{\boldsymbol{e}}_j \tag{7.39}$$

is a complex *polarisation vector* for a given plane wave solution. The requirement that the wave be transverse then amounts to the condition

$$\boldsymbol{e}\cdot\boldsymbol{n} = 0, \tag{7.40}$$

where $\boldsymbol{n}$ is a vector along the direction of travel of the wave. Note that this statement remains generally true, no matter what the direction of the wave: it is always true that the electric field lies in a plane transverse to the direction of travel. The fact that there are two contributions in Eq. (7.39) simply follows from the fact that we need two independent vectors to span a 2D space. Physically, this means that an electromagnetic wave has two independent degrees of freedom associated with it,[4] and we often talk about these as being different *polarisation states.* Above, we have chosen components of the electric field in the $x$ and $y$ directions expressed by $\hat{\boldsymbol{e}}_1$ and $\hat{\boldsymbol{e}}_2$ to be our independent polarisation states, but we could have chosen any independent linear combinations of these.

Let us now look at what the electric field looks like for various choices of the parameters $\{A_i\}$ and $\phi_i$. Without loss of generality, we can set $\phi_1 = 0$, so that we are only sensitive to the relative difference in phase between the two terms in Eq. (7.37). We can then plot the shape in the transverse

[4] It may look as if we have four degrees of freedom in Eq. (7.39), namely the two amplitudes $\{A_i\}$, and the two phases $\{\phi_i\}$. However, we are then taking a real part, which gets rid of two degrees of freedom.

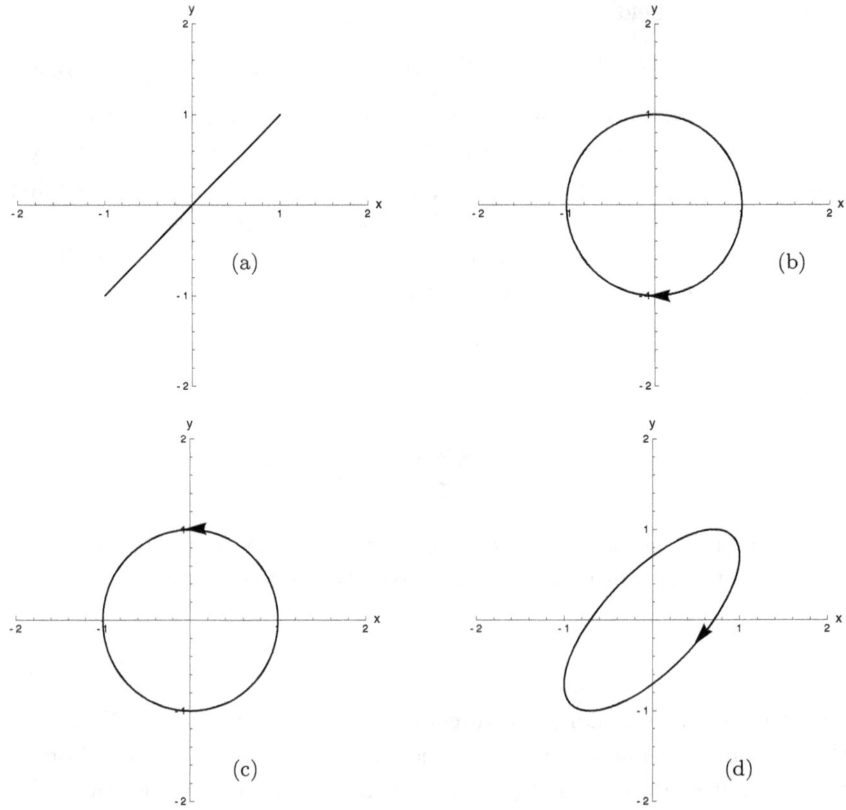


Fig. 7.11 Electric field curves in the transverse plane for $A_1 = A_2 = 1$ and: $(\phi_1, \phi_2) = (0, 0)$ (a); $(\phi_1, \phi_2) = (0, \pi/2)$ (b); $(\phi_1, \phi_2) = (0, -\pi/2)$ (c); $(\phi_1, \phi_2) = (0, \pi/4)$ (d). For the circular and elliptic cases, we show the direction in which the curve is traced out.

plane that the electric field vector makes as time goes on, analogous to the case of Figure 7.10. Note that we show only the boundary of this shape, corresponding to the maximum distance from the origin that the electric field makes in each direction as it oscillates. The first example has non-zero amplitudes in both the $x$ and $y$ directions, and zero relative phase difference. This leads to a straight line as in Figure 7.10, and we may obtain an arbitrary angle by varying the amplitudes $A_i$. The next-most complicated example is that of equal amplitudes and a relative phase difference of $\pm\pi/2$, shown in Figures 7.11(b) and 7.11(c). This gives a circular curve, where the direction of travel of the electric field vector as time progresses is depicted by an arrow. For obvious reasons, the cases of zero or $\pm\pi/2$ phase difference are referred to as *linearly* and *circularly polarised light.* For the latter case, a phase difference of $\pm\pi/2$ yields *left-* and *right-circularly* polarised light, respectively. To understand the names, imagine place the thumb of the relevant hand in the direction of travel of the wave (n.b. this is coming out of the page in our example). Then your fingers will curl in the direction that the electric field moves in, for a wave of the appropriate handedness.

The lower-right panel of Eq. (7.11) shows the case of a phase difference of $\pi/4$, for which the electric field becomes an ellipse, whose semi-major axis is aligned at an angle $\pi/4$ to the $x$-axis. More generally, we can obtain an ellipse aligned at any angle, and with any ratio of semi-major and semi-minor axes, by varying the amplitudes $\{A_i\}$. Thus, so-called *elliptic polarisation* is the most general case for an electromagnetic wave, where circular and linear polarisation emerge as special cases.

In this chapter, we have arrived at Maxwell's equations in detail, and considered one of their most striking consequences, namely the existence of electromagnetic waves. For many purposes, the formulation of electromagnetism that we have given in this chapter is adequate. Alternative formalisms exist, however, that can be much more convenient — particularly if we want to understand the underlying structure of the theory, and where it ultimately comes from. This is the subject of the following chapters.

## Exercises

(1) Verify the identity given in Eq. (7.15).

(2) Show that the vacuum Maxwell equations are symmetric (do not change) under the transformation

$$\boldsymbol{E} \to -c\boldsymbol{B}, \quad \boldsymbol{B} \to \frac{\boldsymbol{E}}{c},$$

known as a *duality transformation*. What breaks this symmetry in the non-vacuum case?

(3) A *quarter-wave plate* is a uniform slab of material that can be used to impose a phase-difference of $\pi/2$ between the $x$ and $y$ components of the electric field of an EM wave travelling in the positive $z$-direction. Explain how this can be used to produce circularly polarised light, from a linearly polarised beam.

# Chapter 8

# Relativity and Maxwell's Equations

We have seen throughout this book that knowledge of Special Relativity (SR) allows us to understand various aspects of electromagnetism. For example, the fact that the electric and magnetic phenomena must unify into a single theory is a consequence of the fact that whether or not charges are moving depends on the observer. Historically, however, the discovery of electromagnetism — and the complete description offered by the Maxwell equations — preceded the development of SR. It is easy to see why, upon recalling the two postulates upon which SR is based:

(1) *Physics is the same in all inertial frames.* An inertial frame is a system of coordinates that is moving at constant speed (relative to all other inertial frames).
(2) *The speed of light $c$ is the same for all observers.*

The first postulate is not unique to SR, and indeed occurs in Newtonian mechanics, when it is usually referred to as *the principle of Galilean Relativity.* However, the second postulate is in marked contrast to the laws of Newtonian mechanics, which state that if an object is moving at velocity $\boldsymbol{v}$ in some inertial frame $S'$, its velocity in a second inertial frame $S$ will be given by simply adding the velocity of the object in $S'$ to the velocity of $S'$ relative to $S$. An electromagnetic wave would then have *different* speeds in different frames, which is not what is implied by the second postulate of SR. We can thus view SR — and all of its weird and wonderful consequences — as being due to the second postulate. It is precisely this postulate that is implied by Maxwell's equations, and which agrees with all experiments that have been carried out to test this idea.

If Maxwell's equations imply SR, it must be possible to cast the equations of electromagnetism in a form such that the full structure of SR is made manifest. That this is not the case in Eqs. (7.16)–(7.19) follows from the fact that these explicitly involve *separate* electric and magnetic fields $\boldsymbol{E}$ and $\boldsymbol{B}$. We know that transforming to a different inertial frame will mix these up in general, and thus the question naturally arises as to whether one can combine the electric and magnetic aspects of the theory into a *single* mathematical object. This is indeed possible, and to work towards it, we will start by reviewing some aspects of Special Relativity.

## 8.1 Relativistic Kinematics

Introducing SR would take an entire textbook by itself. Thus, we will assume that you are familiar with the basics here, and merely provide a brief review of those concepts that are needed most for what follows. Let us first consider two inertial frames $S$ (coordinates $(\boldsymbol{x}, t)$) and $S'$ (coordinates $(\boldsymbol{x}', t')$), and two points $A$ and $B$. The latter will be separated by some distance

$$\Delta \boldsymbol{x} = \boldsymbol{x}_B - \boldsymbol{x}_A$$

in frame $S$, and

$$\Delta \boldsymbol{x}' = \boldsymbol{x}'_B - \boldsymbol{x}'_A$$

in frame $S'$. Now imagine that we set off a light flash from point $A$, which reaches $B$ at some later time. In frames $S$ and $S'$, this takes time $\Delta t$, and $\Delta t'$, respectively. However, by the second postulate above, we must have that

$$\frac{(\Delta \boldsymbol{x})^2}{(\Delta t)^2} = c^2 = \frac{(\Delta \boldsymbol{x}')^2}{(\Delta t')^2}.$$

We therefore arrive at the conclusion that

$$c^2(\Delta t)^2 - \Delta \boldsymbol{x} \cdot \Delta \boldsymbol{x} \tag{8.1}$$

is the same in all frames, and for this reason it is known as the *invariant (spacetime) distance*. We can go further than this in saying how the spatial coordinates and time variable change when we go from one inertial frame to another. These are the well-known *Lorentz transformations*, with which you should already be familiar. They consist of rotations and translations (which are also present in non-relativistic mechanics). In addition, there

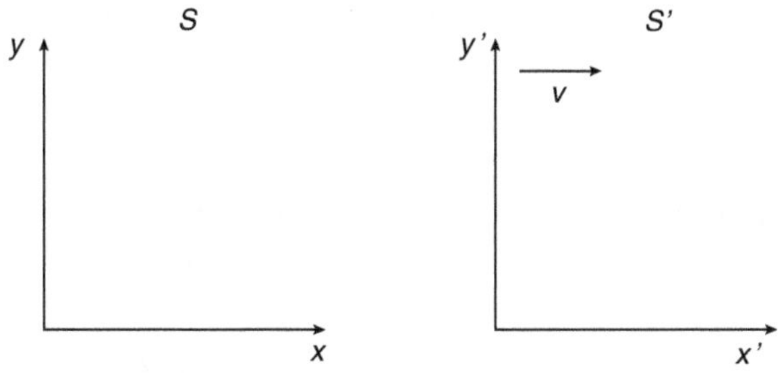


Fig. 8.1 Example of an inertial frame $S'$ which is boosted with respect to an inertial frame $S$.

are *boosts*, corresponding to moving to an inertial frame with a different velocity. To be more specific, let us consider the canonical example of a frame $S'$ that is moving with speed $v$ along the positive $x$-direction relative to some frame $S$, such that the frames coincide at time $t = 0$. This set-up is shown in Figure 8.1, and the Lorentz transformation is in this case given by

$$\begin{aligned} ct' &= \gamma\left(ct - \frac{vx}{c}\right), \\ x' &= \gamma(x - vt), \\ y' &= y, \\ z' &= z, \end{aligned} \tag{8.2}$$

where

$$\gamma = \frac{1}{\sqrt{1 - \frac{v^2}{c^2}}}. \tag{8.3}$$

There are a number of interesting physical effects that arise from these transformation formulae, such as that length and time differences depend on which frame one measures them in. To this end, it is conventional to define the *proper length* $L$ of an object, which corresponds to the length an observer would measure if in a frame at which the object is at rest. Likewise, one can define *proper time* $\tau$ as a difference in time measured by an observer in the rest frame of an object. In other frames, the proper time gets *dilated*, whereas proper length gets *contracted* (see the exercises).

In SR, as in non-relativistic mechanics, we can define energy and momentum, both of which are conserved. We will see a simple way to derive these in the following section, but for now let us just recall that the relativistic momentum of a particle is given by

$$\boldsymbol{p} = \gamma m \boldsymbol{v}, \tag{8.4}$$

where $m$ is the *rest mass* (mass measured in the rest frame of the particle). For speeds $v$ much less than $c$, we may expand the Lorentz factor according to

$$\gamma = \left(1 - \frac{v^2}{c^2}\right)^{-\frac{1}{2}} = 1 + \frac{v^2}{2c^2} + \mathcal{O}\left(\frac{v^4}{c^4}\right). \tag{8.5}$$

Then the relativistic momentum tends to its non-relativistic counterpart ($\boldsymbol{p} = m\boldsymbol{v}$) as required. The total relativistic energy is given by

$$E = \gamma mc^2. \tag{8.6}$$

Again using the expansion of Eq. (8.5), one finds that for low speeds this reduces to

$$E = mc^2 + \frac{mv^2}{2} + \mathcal{O}\left(\frac{v^4}{c^2}\right). \tag{8.7}$$

We see that this consists of the non-relativistic kinetic energy, plus an extra contribution that persists even if the particle is at rest:

$$E_{\text{rest}} = mc^2. \tag{8.8}$$

This is Einstein's famous formula telling us that energy and mass are related to each other. More generally, from the above definitions of relativistic momentum and energy (Eqs. (8.4) and (8.6)), one may show that:

$$E^2 - \boldsymbol{p}^2c^2 = m^2c^4. \tag{8.9}$$

Given that we have not specified the speed of the particle, this relation must be true in all inertial frames. Indeed, the right-hand side of Eq. (8.9) is what results upon explicitly evaluating the left-hand side in a frame in which the particle is at rest ($\boldsymbol{p} = \boldsymbol{0}$, and $E = mc^2$).

## 8.2 Four Vectors

Above, we have written down the Lorentz transformations relating time and space coordinates. Given that space and time mix under such transformations, however, it no longer becomes natural to regard the time variable $t$ (in some frame) as being completely separate from the 3D spatial position $\boldsymbol{x}$.

Instead, we can combine them into a single object known as a *four-vector*, which provides a highly efficient language for discussing all known relativistic theories. Before doing so, let us remind ourselves of what happens in 3D space. As described in Chapter 2, we use *vectors* to describe quantities which have a magnitude and a direction, such as positions and velocities. Given two vectors $\boldsymbol{a}$ and $\boldsymbol{b}$, we can define the dot product of Eq. (2.9), which in Cartesian components takes the form of Eq. (2.15). We can make the latter look a lot more efficient by introducing an alternative notation for the components of a vector: instead of writing $x$, $y$ and $z$ components, we may instead label these by 1, 2, and 3, such that

$$(a_1, a_2, a_3) = (a_x, a_y, a_z). \tag{8.10}$$

A given component of $\boldsymbol{a}$ can then be written as $a_i$ for some $i \in \{1, 2, 3\}$, and we call this the *index notation* for the vector, given that an integer index $i$ is involved! Our Cartesian dot product of Eq. (2.15) can then be recognised as

$$\boldsymbol{a} \cdot \boldsymbol{b} = \sum_{i=1}^{3} a_i b_i. \tag{8.11}$$

As also discussed in Chapter 2, once we have this dot product, we can define the magnitude of a vector via

$$|\boldsymbol{a}| = \sqrt{\boldsymbol{a} \cdot \boldsymbol{a}}. \tag{8.12}$$

We often write $\boldsymbol{a}^2$ to mean $\boldsymbol{a} \cdot \boldsymbol{a}$. Furthermore, if we think of $\boldsymbol{a}$ as a column vector, then $\boldsymbol{a}^T$ (the transpose of $\boldsymbol{a}$) is a row vector, and we can write the dot product in the matrix notation

$$\boldsymbol{a} \cdot \boldsymbol{b} = \boldsymbol{a}^T \boldsymbol{b}, \tag{8.13}$$

which upon multiplying out the vectors is exactly equivalent to Eq. (8.11).

A key property of $|\boldsymbol{a}|$ is that it is preserved under rotations. Let us take the example of a position vector $\boldsymbol{x}$. Under rotations, this becomes some other vector:

$$\boldsymbol{x} \to \boldsymbol{x}' = \mathsf{R}\,\boldsymbol{x}, \tag{8.14}$$

where $\mathsf{R}$ is a rotation matrix. Taking the transpose of this equation gives

$$\boldsymbol{x}^T \to \boldsymbol{x}^T\,\mathsf{R}^T, \tag{8.15}$$

so that the squared magnitude of $\boldsymbol{x}$ transforms as follows:

$$\boldsymbol{x}^2 = \boldsymbol{x}^T\,\boldsymbol{x} \to \boldsymbol{x}^T\,\mathsf{R}^T\,\mathsf{R}\boldsymbol{x}. \tag{8.16}$$

At this point, we may use the fact that rotation matrices are orthogonal, i.e.

$$\mathsf{R}^T\mathsf{R} = \mathsf{I}, \tag{8.17}$$

where $\mathsf{I}$ is the identity matrix. We thus find

$$\boldsymbol{x}^T\,\boldsymbol{x} \to \boldsymbol{x}^T\,\boldsymbol{x}, \tag{8.18}$$

namely that rotations do not change the magnitude of $\boldsymbol{x}$, as stated above. If you have not seen rotation matrices before, don't worry: we can instead use geometric intuition to tell us that the length of a vector is preserved under rotations. We may think of a position vector as being an arrow in space, pointing from the origin to a point of interest. Rotating this arrow clearly does not change its length! Indeed, we can go further than this and say that the dot product of any two vectors $\boldsymbol{a}\cdot\boldsymbol{b}$ is preserved by rotations. The mathematical proof is very similar to that given above, and again has a straightforward geometric interpretation: the dot product of $\boldsymbol{a}$ and $\boldsymbol{b}$ corresponds to the component of $\boldsymbol{a}$ that is in the direction of $\boldsymbol{b}$. Rotating the pair of vectors (viewed as arrows in space) does not change this quantity. Note that, in conventional 3D space, rotations mix space dimensions with other space dimensions — time is completely separate, and remains unaffected.

As already hinted at above, the concept of four vectors comes from the observation that Lorentz transformations mix up space and time. We can see this explicitly in the Lorentz transformation of Eq. (8.2), which describes a boost. This suggests that we should not consider space and time to be separate. Rather, we should combine them into a single entity: *spacetime*. The spacetime we live in is 4D (as far as we can tell): namely, three dimensions of space and one of time. Given the concept of spacetime, we can define analogues of vectors and dot products. A vector in spacetime is called a 4-vector (where the 4 comes from the dimension of spacetime), and the simplest 4-vector to define is *position*:

$$x = \begin{pmatrix} ct \\ x \\ y \\ z \end{pmatrix}. \tag{8.19}$$

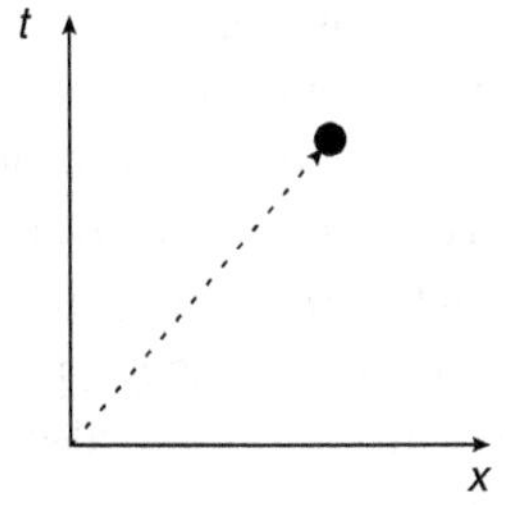


Fig. 8.2 A 4-vector $x$ can be thought of as an arrow pointing from the origin of spacetime to the location of an event, where only one dimension of space is shown.

Note that I have here used the common convention for 4-vectors that they are written as simple symbols with no underline or arrow (this avoids confusion with 3-vector). Unfortunately, in this case this leads to an additional confusion, in that the symbol $x$ appears on both sides of Eq. (8.19): on the left, it stands for the whole 4-vector. On the right, it appears as the position component associated with the $x$-axis! Meanings should hopefully be clear from the context. This 4-vector denotes the location of a point in spacetime. We normally call such points *events*, as they correspond to something at a definitive point in space, and at a particular time. As in normal 3D space, we can think of this vector as pointing from the origin to the location of the event in a spacetime diagram, as in Figure 8.2.

To make a further analogy with the components of a vector in ordinary space, it is inconvenient to keep labelling the components as $ct$, and $x$, $y$ etc. Let us instead define

$$x^0 = ct, \quad x^1 = x, \quad x^2 = y, \quad x^3 = z,$$

where we have adopted a widely used, but rather unfortunate, notation: $x^2$ means the $y$-like component of the 4-vector, and not the $x$-component squared! However, given this notation, we may write a generic component of the 4-vector as

$$x^\mu, \quad 0 \le \mu \le 3.$$

This is another convention — Greek letters are commonly used to represent the components of 4-vectors. It reduces confusion with the components of ordinary 3D vectors, for which we tend to use Latin indices (e.g. $a_i$, $b_j$). Also, you may see slightly different choices in textbooks and the research literature. Here we have chosen the "zeroth" component to involve the time, but we could have said that this is the "fourth" component. Such choices are clearly irrelevant for physics, but one must be self-consistent.

Just as for 3D vectors, we can represent other types of quantity by 4-vectors, as well as position. However, it is conventional to refer to the zeroth component of any 4-vector as the *time-like* component, and the other components (1,2,3) as the *space-like* components, by analogy with the position 4-vector. Given the above definition, we can also look at defining operations involving 4-vectors, which are very similar to their 3-vector counterparts. First, there is addition of 4-vectors, given by

$$(x+y)^\mu = x^\mu + y^\mu. \tag{8.20}$$

In matrix notation, this reads

$$\begin{pmatrix} x^0 \\ x^1 \\ x^2 \\ x^3 \end{pmatrix} + \begin{pmatrix} y^0 \\ y^1 \\ y^2 \\ y^3 \end{pmatrix} = \begin{pmatrix} x^0 + y^0 \\ x^1 + y^1 \\ x^2 + y^2 \\ x^3 + y^3 \end{pmatrix}, \tag{8.21}$$

which is a very natural thing to define! We can also define multiplication by a number $\lambda$ (which could be complex in general), by

$$(\lambda x)^\mu = \lambda x^\mu \tag{8.22}$$

or, in matrix notation,

$$\lambda \begin{pmatrix} x^0 \\ x^1 \\ x^2 \\ x^3 \end{pmatrix} = \begin{pmatrix} \lambda x^0 \\ \lambda x^1 \\ \lambda x^2 \\ \lambda x^3 \end{pmatrix}. \tag{8.23}$$

How do 4-vectors transform between inertial frames? We know the answer to this already — they obey Lorentz transformations. For example, we can rewrite Eq. (8.2) using 4-vector and matrix notation, as

$$\begin{pmatrix} x'^0 \\ x'^1 \\ x'^2 \\ x'^3 \end{pmatrix} = \begin{pmatrix} \gamma & -\gamma\beta & 0 & 0 \\ -\gamma\beta & \gamma & 0 & 0 \\ 0 & 0 & 1 & 0 \\ 0 & 0 & 0 & 1 \end{pmatrix} \begin{pmatrix} x^0 \\ x^1 \\ x^2 \\ x^3 \end{pmatrix}, \tag{8.24}$$

where we defined

$$\beta = \frac{v}{c}, \quad \gamma = (1-\beta^2)^{-\frac{1}{2}} \tag{8.25}$$

(the latter is the usual Lorentz factor). This looks remarkably like the transformation of 3-vectors under rotations, and indeed the mathematics here is very similar. We can thus carry forward lots of intuition about how 3-vectors transform under rotations to the present context — 4-vectors transforming under Lorentz transformations. There are, however, important differences, which can ultimately be traced to the fact that a time dimension is intrinsically different to a space dimension. We will see this in Section 8.3.

Note that one can rewrite the above using our index notation as

$$x'^{\mu} = \sum_{\nu=0}^{3} \Lambda^{\mu}{}_{\nu} x^{\nu}, \tag{8.26}$$

where $\Lambda^{\mu}{}_{\nu}$ is the component of the Lorentz transformation matrix in the $\mu$th row and $\nu$th column. You may wonder why the $\nu$ index has been written as a lower index, whereas the others are upper indices. This will hopefully become clear in what follows.

## 8.3 The Dot Product for 4-Vectors

In the previous section, we saw that 4-vectors transforming under Lorentz transformations look very like 3-vectors transforming under rotations, at least as far as the matrix notation is concerned. We also saw that rotations in the 3-vector case preserved the magnitude of 3-vectors. Is there an analogue of this for 4-vectors? In other words, is there a dot product for 4-vectors that is preserved by Lorentz transformations? The answer is yes, and the dot product for 4-vectors is defined to be (in Cartesian coordinates)

$$x \cdot y = x^0 y^0 - x^1 y^1 - x^2 y^2 - x^3 y^3. \tag{8.27}$$

Defining the "magnitude" of a 4-vector $x$ by $|x| = \sqrt{x \cdot x}$ (by analogy with the 3-vector case), one finds

$$|x|^2 = (x^0)^2 - \boldsymbol{x}^2 \equiv c^2 t^2 - \boldsymbol{x}^2, \tag{8.28}$$

where $\boldsymbol{x}$ denotes the three space-like components of the 4-vector $x$ (i.e. these are a conventional 3-vector). Given that $\boldsymbol{x}$ and $t$ denote the spatial and temporal displacements from the origin of spacetime, we see that Eq. (8.28) is the invariant distance that we derived earlier, and thus is certainly preserved by Lorentz transformations. For a more general argument, one may first rewrite the 4-vector dot product in a more convenient form.

Let us introduce the vector

$$x_\mu = (x^0, -\boldsymbol{x}), \tag{8.29}$$

related by $x^\mu$ by the fact that the space-like components have been reversed. These are known as the *covariant* components of the vector, rather than the original *contravariant* components $x^\mu$. There is no confusion in principle with this notation, as we have now labelled the $\mu$ index as being downstairs, rather than upstairs. We can then write the 4-vector dot product in index notation as

$$x \cdot y = \sum_{\mu=0}^{3} x_\mu \, y^\mu. \tag{8.30}$$

From Eq. (7.9), we may surmise that

$$\begin{pmatrix} x'^0 \\ -x'^1 \\ -x'^2 \\ -x'^3 \end{pmatrix} = \begin{pmatrix} \gamma & +\gamma\beta & 0 & 0 \\ +\gamma\beta & \gamma & 0 & 0 \\ 0 & 0 & 1 & 0 \\ 0 & 0 & 0 & 1 \end{pmatrix} \begin{pmatrix} x^0 \\ -x^1 \\ -x^2 \\ -x^3 \end{pmatrix}, \tag{8.31}$$

where the matrix in this case has $\beta \to -\beta$ relative to Eq. (5.2). This corresponds to setting $v \to -v$, such that the matrix in Eq. (8.31) is the *inverse* of the Lorentz transformation matrix appearing in Eq. (7.9). We thus see that the 4-vector with a downstairs index transforms according to the inverse of the matrix that transforms the vector with an upstairs index. In index notation (to be compared with Eq. (8.26)):

$$x'_\mu = \sum_\nu x_\nu (\Lambda^{-1})^\nu{}_\mu. \tag{8.32}$$

Then the dot product of two vectors transforms as

$$\begin{aligned} x' \cdot y' &= \sum_{\mu,\nu,\alpha} x_\mu (\Lambda^{-1})^\mu{}_\nu \, \Lambda^\nu{}_\alpha \, y^\alpha \\ &= \sum_\mu x_\mu \, y^\mu \\ &= x \cdot y. \end{aligned} \tag{8.33}$$

In the second line, we have used the fact that the inverse Lorentz transformation matrix acts on the Lorentz matrix itself to give the identity matrix or, in index notation,

$$(\Lambda^{-1})^{\mu}{}_{\nu}\,\Lambda^{\nu}{}_{\alpha} = \delta^{\mu}_{\alpha}, \tag{8.34}$$

where $\delta^{\mu}_{\alpha}$ is the Kronecker symbol (1 if $\mu = \alpha$, otherwise zero). We therefore conclude that the dot product $x \cdot y$ of 4-vectors is Lorentz invariant.

This is an amazingly useful thing to have in mind. Whenever we see a dot product of any two 4-vectors, we can evaluate it in any Lorentz frame we choose, and the answer will be the same. There may then be frames in which the answer is much simpler to calculate! Another useful property to note is that we can equally write the dot product of Eq. (8.30) as

$$x \cdot y = \sum_{\mu=0}^{3} x^{\mu} y_{\mu}, \tag{8.35}$$

i.e. where the index is lowered on $y^{\mu}$ instead of $x^{\mu}$. This follows straightforwardly from Eq. (8.29), if we apply it to $y^{\mu}$.

## 8.4 The Metric Tensor

In the previous section, we wrote the dot product in terms of two different sets of 4-vector components: $x^{\mu}$ represented the physical components of the position, and $x_{\mu}$ denoted the 4-vector with its space-like components reversed. Can we write the dot product for 4-vectors using only the original position vectors? This is indeed possible, if we introduce a matrix known as the *metric tensor*

$$\eta_{\mu\nu} = \begin{pmatrix} 1 & 0 & 0 & 0 \\ 0 & -1 & 0 & 0 \\ 0 & 0 & -1 & 0 \\ 0 & 0 & 0 & -1 \end{pmatrix} \tag{8.36}$$

(i.e. $\eta_{\mu\nu}$ is the component in the $\mu$th row and $\nu$th column). The dot product can then be written

$$x \cdot y = \sum_{\mu,\nu} \eta_{\mu\nu}\, x^{\mu}\, y^{\nu}, \tag{8.37}$$

as can checked by explicit calculation, after reinterpreting this formula in matrix notation (which means lining up the indices appropriately):

$$x \cdot y = \begin{pmatrix} x^0 & x^1 & x^2 & x^3 \end{pmatrix} \begin{pmatrix} 1 & 0 & 0 & 0 \\ 0 & -1 & 0 & 0 \\ 0 & 0 & -1 & 0 \\ 0 & 0 & 0 & -1 \end{pmatrix} \begin{pmatrix} y^0 \\ y^1 \\ y^2 \\ y^3 \end{pmatrix}. \tag{8.38}$$

We can compare this with the dot product for 3-vectors in ordinary space, which we could write as

$$\boldsymbol{x} \cdot \boldsymbol{y} = \sum_{i=1}^{3} \eta_{ij} x_i y_j, \tag{8.39}$$

where

$$\eta_{ij} = \begin{pmatrix} 1 & 0 & 0 \\ 0 & 1 & 0 \\ 0 & 0 & 1 \end{pmatrix}. \tag{8.40}$$

The reason we do not usually bother doing this is that in this case $\eta_{ij}$ is imply the identity matrix, and so doesn't really do anything. Nevertheless, it suggests that we have found two specific examples of dot products, and that the correct way to define a dot product is to sandwich two vectors with a metric tensor. Different metric tensors give us different types of mathematical space.

The case of ordinary 3-vectors (or indeed any space in which the metric is the identity) is called *Euclidean space*, as this is the type of space that was studied by Euclid in ancient Greece. The case of Eq. (8.36), that describes the 4D spacetime of special relativity, is called *Minkowski space*, after the mathematician that first developed this language in that context.

There is another use for the metric tensor, aside from forming dot products. From Eq. (8.37), we can surmise that

$$y_\mu = \eta_{\mu\nu} y^\nu, \tag{8.41}$$

which is easy to check by explicit calculation. That is,

$$\eta_{\mu\nu} y^\nu = \begin{pmatrix} 1 & 0 & 0 & 0 \\ 0 & -1 & 0 & 0 \\ 0 & 0 & -1 & 0 \\ 0 & 0 & 0 & -1 \end{pmatrix} \begin{pmatrix} y^0 \\ y^1 \\ y^2 \\ y^3 \end{pmatrix} = \begin{pmatrix} y^0 \\ -y^1 \\ -y^2 \\ -y^3 \end{pmatrix},$$

agreeing with our rule for lowering an index. Thus, the metric tensor can be used to lower indices of 4-vector and related quantities to be discussed below. Likewise, we can define a matrix

$$\eta^{\mu\nu} = \begin{pmatrix} 1 & 0 & 0 & 0 \\ 0 & -1 & 0 & 0 \\ 0 & 0 & -1 & 0 \\ 0 & 0 & 0 & -1 \end{pmatrix}, \tag{8.42}$$

which can be used to *raise* indices as follows:

$$y^{\mu} = \eta^{\mu\nu} y_{\nu}, \tag{8.43}$$

as may again be checked. It may seem strange to define this much notation, especially given that the matrix components of $\eta^{\mu\nu}$ and $\eta_{\mu\nu}$, as written here, are the same. However, we have worked in Cartesian coordinates throughout, and it just so happens that in other coordinate systems, $\eta^{\mu\nu}$ and $\eta_{\mu\nu}$ can be quite different. What is *always* true, however, is that they are inverse matrices of each other:

$$\eta^{\mu\nu}\eta_{\nu\alpha} = \delta^{\mu}_{\alpha}, \tag{8.44}$$

which follows from the fact that if we lower and then raise the index of a 4-vector, we are back to what we started with!

## 8.5 The Velocity 4-Vector

In ordinary 3-space, once we have position vectors we can define velocities via

$$\boldsymbol{v} = \frac{d\boldsymbol{x}}{dt}. \tag{8.45}$$

Similarly, given a 4-vector $x^{\mu}$, we can define the *velocity 4-vector* of a particle:

$$u^{\mu} = \frac{dx^{\mu}}{d\tau}, \tag{8.46}$$

where $\tau$ is the proper time measured in a frame in which the particle is at rest. The reason we have used the proper time rather than the time itself is that the meaning of the proper time is the same in all frames (i.e. it is defined to be the time measured in the rest frame). The derivative in

Eq. (8.46) can be defined using a limiting procedure in which one divides the 4-vector $dx^\mu$ by a small proper time difference $d\tau$, as $d\tau \to 0$. Division by an invariant number does not change the property of being a 4-vector, so that $u^\mu$ is indeed a 4-vector as desired. This also means that if we take a dot-product of the 4-velocity with any other 4-vector, we will get a Lorentz invariant. Let us stress, though, that the velocity 4-vector is not itself the same in all frames, as $x^\mu$ is frame-dependent.

We can write the 4-velocity in terms of the time coordinate in the frame of interest by recalling that proper time gets dilated according to

$$dt = \gamma d\tau \quad \Rightarrow \quad \frac{d}{d\tau} = \gamma \frac{d}{dt}. \tag{8.47}$$

From Eq. (8.19), we see that the 4-velocity has components

$$u^\mu = (\gamma c, \gamma \boldsymbol{v}), \tag{8.48}$$

where $\boldsymbol{v}$ is the usual 3-velocity. Using the 4-vector dot product, we then find that the (squared) magnitude of $u$ is

$$\begin{aligned} u^2 &= \sum_\mu u_\mu \, u^\mu \\ &= \gamma^2 c^2 \left(1 - \frac{v^2}{c^2}\right) \\ &= c^2. \end{aligned} \tag{8.49}$$

This is clearly Lorentz invariant, as required, as $c$ has the same value in all frames. Indeed, the right-hand side is what we get if we explicitly evalulate the 4-velocity in the rest frame of the particle: it then has components $(c, \mathbf{0})$.

Given the 4-velocity, we can then define a 4-momentum by multiplying by the rest mass $m$:

$$p^\mu = m u^\mu = (\gamma m c, \gamma m \boldsymbol{v}). \tag{8.50}$$

Given that $u^\mu$ is a 4-vector, and the rest mass is the same in all frames (i.e. it is always defined as the mass in a particular frame), we see that $p^\mu$ is indeed a 4-vector. Also, it is very easy to remember this definition, as it is extremely similar to the usual definition of momentum in non-relativistic mechanics (except that we now have 4-vectors). Using Eq. (8.48), the components $p^\mu$ are found to be

$$\begin{aligned} p^\mu &= (\gamma mc, \gamma m\boldsymbol{v}) \\ &= (E/c, \boldsymbol{p}), \end{aligned} \tag{8.51}$$

where we have recognised the relativistic energy and momentum of Eqs. (3.6) and (8.4) in the second line. The above 4-vector language has given a very elegant way to rederive this. It does more than this though — it tells us that energy and momentum (up to factors of $c$) form a 4-vector. Thus, they must mix up under Lorentz transformations in a similar way to time and space. In the 4-vector notation, this reads

$$p'^\mu = \sum_\nu \Lambda^\mu{}_\nu \, p^\nu. \tag{8.52}$$

Also, dot products of any two 4-momenta (e.g. $p_1 \cdot p_2$) must be Lorentz invariant, as this is true for all dot products of 4-vectors. In particular this applies to

$$p^2 = \sum_\mu p_\mu \, p^\mu = \frac{E^2}{c^2} - \boldsymbol{p}^2. \tag{8.53}$$

In the frame in which a given particle is at rest, its 4-momentum is given by

$$p^\mu = (mc, \mathbf{0}) \quad \Rightarrow \quad p^2 = m^2c^2.$$

However, if $p^2$ is Lorentz invariant, this must be the value in all frames, so that

$$\frac{E^2}{c^2} - \boldsymbol{p}^2 = m^2c^2 \quad \Rightarrow \quad E^2 - \boldsymbol{p}^2c^2 = m^2c^4. \tag{8.54}$$

This is the same relation between momentum and energy that we have found before, but derived in a much more elegant language.

## 8.6 Tensors and Relativistic Equations

We have now seen various examples of 4-vectors, where what unites them all is how they transform under Lorentz transformations. We will also see examples of more complicated mathematical objects, which have a number of upstairs or downstairs spacetime indices in general. The name for such objects is *tensors* and they by definition transform between frames by extending Eqs. (8.26) and (8.32) to each upstairs or downstairs index, respectively:

$$T'^{\mu_1\ldots\mu_n}_{\nu_1\ldots\nu_m} = \Lambda^{\mu_1}{}_{\alpha_1}\cdots\Lambda^{\mu_n}{}_{\alpha_n}(\Lambda^{-1})^{\beta_1}{}_{\nu_1}\cdots(\Lambda^{-1})^{\beta_m}{}_{\nu_m}T^{\alpha_1\ldots\alpha_n}_{\beta_1\ldots\beta_m}. \tag{8.55}$$

That is, the components of the tensor in a frame $S'$ can be obtained from those in a frame $S$ by acting appropriately with the Lorentz transformation matrix and its inverse. In simplifying the structure of this equation, we have used the *Einstein summation convention*, by which it is understood that any pair of repeated indices is summed over. In other words, we do not bother including explicit sums of the form $\sum_{\alpha_i}$, etc. on the right-hand side of Eq. (8.55).

For a given theory, writing its defining equations in terms of tensors (of which 4-vectors are a special case) means that relativity is made manifest, in that the form of the equations will be the same in any coordinate frame. Consider, for example, an equation of the form

$$A'^{\mu}T'_{\mu\nu} = B'_{\nu}, \tag{8.56}$$

involving quantities defined in some frame $S'$. There are two types of index on the left-hand side: (i) a *dummy index* $\mu$, that is summed over, and thus cannot appear on the right-hand side and (ii) a *free index*, that is not summed over, and must therefore match up on both sides of the equation. Using Eq. (8.55), we can write all quantities in Eq. (8.56) in terms of quantities in another frame $S$. We get

$$\begin{aligned} A'^{\mu}T'_{\mu\nu} &= [\Lambda^{\mu}{}_{\alpha}A^{\alpha}][(\Lambda^{-1})^{\beta}{}_{\mu}(\Lambda^{-1})^{\gamma}{}_{\nu}T_{\beta\gamma}] \\ &= (\Lambda^{-1})^{\beta}{}_{\mu}\Lambda^{\mu}{}_{\alpha}(\Lambda^{-1})^{\gamma}{}_{\nu}A^{\alpha}T_{\beta\gamma}, \end{aligned} \tag{8.57}$$

where in the second line we have simply moved factors around, given that each component of a Lorentz transformation matrix is itself just a number. The first two factors on the left-hand side (remembering the implicit summation over $\mu$) consist of the matrix product of the Lorentz transformation matrix with itself, which gives the identity matrix with indices $\alpha$ and $\beta$:

$$(\Lambda^{-1})^{\beta}{}_{\mu}\Lambda^{\mu}{}_{\alpha} = \delta^{\beta}_{\alpha}, \quad \delta^{\beta}_{\alpha} = \begin{cases} 1, & \alpha = \beta; \\ 0, & \alpha \neq \beta. \end{cases}$$

This then acts on the 4-vector $A^{\alpha}$ to give

$$\delta^{\beta}_{\alpha}A^{\alpha} = A^{\beta}.$$

To see why, note that there is an implicit summation over $\alpha$ from the summation convention. However, in the sum over $\alpha$, only the term with $\alpha = \beta$ gives a non-zero contribution, from the definition of $\delta^\beta_\alpha$. Equation (8.57) then becomes

$$A'^\mu T'_{\mu\nu} = (\Lambda^{-1})^\gamma{}_\nu A^\beta T_{\beta\gamma} = (\Lambda^{-1})^\gamma{}_\nu B_\gamma,$$

where in the final equality we have written the right-hand side of Eq. (8.56) in terms of quantities in $S$. This reveals the presence of a common factor of $(\Lambda^{-1})^\gamma{}_\nu$ on both sides, corresponding to a matrix multiplication by the inverse Lorentz transformation. This could be removed by multiplying both sides of the equation with the Lorentz transformation matrix itself, such that we find

$$A^\beta T_{\beta\gamma} = B_\gamma,$$

which has exactly the same form as the original equation in $S'$. If you are new to manipulations in index notation such as these, the above argument will be very heavy, and you are doing very well in merely keeping up! Careful scrutiny of the above, however, reveals that we were *guaranteed* that the equations would match up in both coordinate frames, given that the free indices have to match up on both sides of any tensor equation.

In summary, any equation written in terms of tensors will have the same form in any inertial frame. Let us then see how to write the equations of electromagnetism in such a form.

## 8.7 The 4-Potential and Field Strength

In finding a tensorial form for Maxwell's equations, we might start by trying to find a single 4-vector quantity that contains both electricity and magnetism. However, we cannot make a 4-vector out of the electric and magnetic fields: the latter have three components each, and thus cannot be used to make a 4-component object. There is, however, a 4-vector that one can make, from which the electric and magnetic fields can both be obtained, and which generalises the notion of the electrostatic potential $V(\boldsymbol{x})$ that we saw in Chapter 3. We saw that we cannot introduce a single-valued scalar potential for the magnetic field, which is non-conservative in general. But that does not mean that we can't have a potential of a more complicated type! To see this, note that Eq. (7.17) tells us that the divergence of the magnetic field is everywhere zero. We also have Eq. (7.15), that tells us that the divergence of the curl of any vector field is automatically zero.

Conversely, any field with zero divergence can be written as the curl of a vector field, so that Eq. (7.17) implies

$$\boldsymbol{B} = \nabla \times \boldsymbol{A}, \tag{8.58}$$

for some $\boldsymbol{A}$. The vector field $\boldsymbol{A}(\boldsymbol{x})$ is known as the *magnetic vector potential*, and is clearly not unique: we may add to $\boldsymbol{A}$ any vector field whose curl vanishes, and get the same magnetic field $\boldsymbol{B}$. Despite its name, $\boldsymbol{A}$ must also enter the expression for the electric field. In Chapter 3, we saw that the electric field is conservative in static situations, and given in terms of the electrostatic potential by Eq. (3.35). However, in time-varying situations, there is a non-conservative component to the electric field, as required by Faraday's law. Consistency with Eq. (7.18) then uniquely fixes

$$\boldsymbol{E} = -\nabla V(\boldsymbol{x}) - \frac{\partial \boldsymbol{A}}{\partial t}. \tag{8.59}$$

To see this, note that upon taking the curl of $\boldsymbol{E}$, we may use the identity

$$\nabla \times (\nabla f(\boldsymbol{x})) = 0, \tag{8.60}$$

valid for *any* scalar function $f(\boldsymbol{x})$. Thus, the first term on the right-hand side of Eq. (8.59) vanishes if we take the curl, leaving

$$\nabla \times \boldsymbol{E} = -\nabla \times \left(\frac{\partial \boldsymbol{A}}{\partial t}\right) = -\frac{\partial}{\partial t}(\nabla \times \boldsymbol{A}), \tag{8.61}$$

where we have used the fact that the order in which we perform space or time derivatives does not matter. Using Eq. (8.58) leads to Eq. (7.18) as required.

The electrostatic potential has one (scalar) degree of freedom, and the magnetic vector potential has three. Given that they are both potentials, we can then ask if they can be put together to make a 4-vector. One can indeed do this, and the correct combination turns out to be

$$A^{\mu} = \left(\frac{V}{c}, \boldsymbol{A}\right), \tag{8.62}$$

where the inverse factor of $c$ in the zeroth component is needed for dimensional reasons. This is called the 4-potential, and implies that we may also define a quantity with a downstairs index as

$$A_\mu = \left(\frac{V}{c}, -\boldsymbol{A}\right). \tag{8.63}$$

It may not be obvious just yet that Eq. (8.62) is a genuine 4-vector, in that it transforms correctly under Lorentz transformations. However, we will be able to justify this statement in what follows. Our aim is to obtain the Maxwell equations from the 4-potential, and to do this, we are going to have to take derivatives of the gauge field, given that space and time derivatives are involved in converting the electrostatic and magnetic vector potentials into the physical electric and magnetic fields (Eqs. (8.58) and (8.59)). In order to cast these derivatives in a properly relativistic language, we must also write them in terms of 4-vectors. To this end, we can define the 4-vector operator

$$\partial_\mu \equiv \frac{\partial}{\partial x^\mu} = \left(\frac{1}{c}\frac{\partial}{\partial t}, \nabla\right). \tag{8.64}$$

This is like a 4D generalisation of the vector operator $\nabla$ that appears in usual 3D vector calculus, and indeed $\nabla$ appears in the spacelike components of Eq. (8.64). Like $\nabla$, Eq. (8.64) is an abstract operator, which is assumed to act on anything that it appears to the left of. To see that Eq. (8.64) is a genuine 4-vector, let us consider transforming to an inertial frame with coordinates $x'^\mu$. From the chain rule, we can write

$$\frac{\partial}{\partial x'^\mu} = \frac{\partial x^\nu}{\partial x'^\mu}\frac{\partial}{\partial x^\nu}.$$

From Eq. (8.26) we find

$$\frac{\partial x'^\mu}{\partial x^\nu} = \Lambda^\mu{}_\nu \quad \Rightarrow \quad \frac{\partial x^\nu}{\partial x'^\mu} = (\Lambda^{-1})^\nu{}_\mu. \tag{8.65}$$

We thus have

$$\partial'_\mu = (\Lambda^{-1})^\nu{}_\mu \partial_\nu,$$

which comparison with Eq. (8.55) shows is the correct transformation law for a 4-vector with a downstairs index. For future reference, it is worth noting that Eq. (8.65) allows us to rewrite Eq. (8.55) as

$$T'^{\mu_1\ldots\mu_n}{}_{\nu_1\ldots\nu_m} = \frac{\partial x'^{\mu_1}}{\partial x^{\alpha_1}}\cdots\frac{\partial x'^{\mu_n}}{\partial x^{\alpha_n}}\cdots\frac{\partial x^{\beta_1}}{\partial x'^{\nu_1}}\cdots\frac{\partial x^{\beta_n}}{\partial x'^{\nu_n}}\cdots T^{\alpha_1\ldots\alpha_n}{}_{\beta_1\ldots\beta_m}. \tag{8.66}$$

Let us now consider the tensor

$$F_{\mu\nu} = \partial_\mu A_\nu - \partial_\nu A_\mu. \tag{8.67}$$

Given each of the indices on the left-hand side can take one of four values, we may think of Eq. (8.67) as defining the components of a $4 \times 4$ matrix:

$$F_{\mu\nu} = \begin{pmatrix} F_{00} & F_{01} & F_{02} & F_{03} \\ F_{10} & F_{11} & F_{12} & F_{13} \\ F_{20} & F_{21} & F_{22} & F_{23} \\ F_{30} & F_{31} & F_{32} & F_{33} \end{pmatrix}, \tag{8.68}$$

where, in a slight abuse of notation, we let the left-hand side denote the set of all possible components. The components of this matrix turn out to be very interesting! First of all, it is straightforward to see from the definition of Eq. (8.67) that

$$F_{\mu\nu} = -F_{\nu\mu}, \tag{8.69}$$

i.e. that the matrix of Eq. (8.68) must be antisymmetric. This implies that the diagonal elements are zero, and that it is sufficient to find the entries in the upper triangle of the matrix in order to fix all of its components. Focusing on the top row, we have

$$\begin{aligned} F_{0i} &= \partial_0 A_i - \partial_i A_0 \\ &= -\frac{1}{c}\frac{\partial A_i}{\partial t} - \nabla_i \left(\frac{V}{c}\right), \end{aligned} \tag{8.70}$$

where we have used Eqs. (8.63) and (8.64). From Eq. (8.59), however, we recognise in Eq. (8.70) the $i$th component of the electric field, and thus one has

$$F_{0i} = -\frac{E_i}{c}. \tag{8.71}$$

For the other components, let us consider

$$F_{12} = \partial_1 A_2 - \partial_2 A_1 = (\nabla \times \boldsymbol{A})_3 = B_3, \tag{8.72}$$

where Eq. (8.58) has been used. Likewise, we find

$$F_{23} = B_1, \quad F_{13} = -B_2, \tag{8.73}$$

so that the full matrix of Eq. (8.68) is seen to be

$$F_{\mu\nu} = \begin{pmatrix} 0 & E_x/c & E_y/c & E_z/c \\ -E_x/c & 0 & -B_z & B_y \\ -E_y/c & B_z & 0 & -B_x \\ -E_z/c & -B_y & B_x & 0 \end{pmatrix}, \tag{8.74}$$

where we have reverted to Cartesian notation for the vector components. We see that the single object $F_{\mu\nu}$ unifies the electric and magnetic fields into a single object. It is called the *field strength tensor*, and will be a key component of the relativistic Maxwell equations. Note that there is a very good reason why an antisymmetric two-index tensor is needed to describe the electric and magnetic fields. If these are to be combined into a single object, then we need this object to have six degrees of freedom, given that it must contain the components of two 3-vectors. A 4-vector is insufficient, as it only has four components. On the other hand, a general two-index tensor in four spacetime dimensions is inappropriate, as it has 16 components. However, an antisymmetric 2-index tensor has precisely 6 components in four spacetime dimensions, and thus is precisely the right sort of quantity to unify $\boldsymbol{E}$ and $\boldsymbol{B}$!

## 8.8 The 4-Vector Current Density

In addition to the electric and magnetic fields, we also need to find a natural relativistic form for the charge and current densities appearing in Eqs. (7.16)–(7.19). Given that $\rho$ is a number, and $\boldsymbol{J}$ a 3-vector, it is natural to try to form a *4-vector current density* as follows:

$$j^\mu = (c\rho, \boldsymbol{J}), \tag{8.75}$$

where the factor of $c$ in the zeroth component is there for dimensional reasons. We can argue that this is correct as follows. First, consider an amount of charge $Q$ at rest in some cubic volume $V$. If the rest-frame of $Q$ is moving with respect to some other frame $S'$, the latter will see the volume $V$ contracted along the direction of motion, to form a volume

$$V' = \frac{V}{\gamma}.$$

Thus, in $S'$ we see a charge density

$$\rho' = \frac{\rho}{V'} = \frac{\gamma\rho}{V} = \gamma\rho.$$

We see that the charge density is increased in the same way that time is, and thus that — if first multiplied by $c$ — it should indeed be regarded as the zeroth component of a 4-vector, rather than a scalar. We can then fix the spacelike components by requiring that charge is conserved in all frames. Consider a fixed volume $V$ in a frame $S$. The total amount of charge in this volume will be

$$Q = \int\int\int_V \rho(\boldsymbol{x}) dV,$$

so that the rate of change of this with time is

$$\frac{dQ}{dt} = \int\int\int_V \frac{\partial \rho(\boldsymbol{x}, t)}{\partial t} dV,$$

where we have used the fact that our volume is fixed to take the partial derivative inside the volume integral. Conservation of charge implies that this rate of change must be related to the rate at which charge is leaving the volume $V$. We can find the latter by taking the surface integral of the current density $\boldsymbol{J}$ over the surface $S$ that bounds the volume $V$ (Figure 8.3), such that

$$\frac{dQ}{dt} = -\int\int_S d\boldsymbol{S} \cdot \boldsymbol{J},$$

where the minus sign is due to the fact that a positive flux of charge corresponds to a decrease of the charge $Q$ inside $V$. Using the divergence theorem, we may rewrite the surface integral, so that we have

$$\int\int\int_V \frac{\partial \rho}{\partial t} dV = -\int\int\int_V \nabla \cdot \boldsymbol{J} \, dV.$$

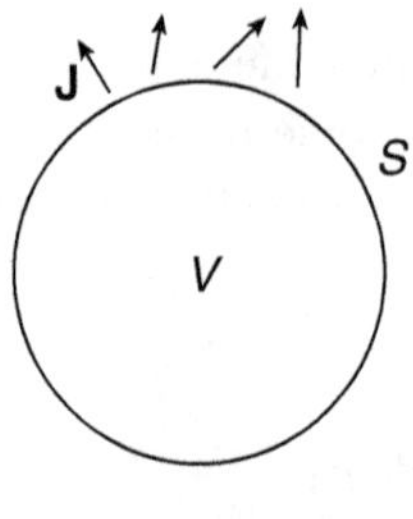


Fig. 8.3 Conservation of charge in a volume $V$ relates the charge leaving the volume to the total charge inside.

Finally, the fact that this condition has to be true for *any* volume $V$ allows us to remove the integral signs, thus finding

$$\frac{\partial \rho}{dt} + \nabla \cdot \boldsymbol{J} = 0. \tag{8.76}$$

This is sometimes called the *continuity equation*, and similar equations crop up throughout physics, when any sort of quantity is conserved. From Eqs. (8.64) and (8.75), we see that this equation may be neatly written as[1]

$$\partial_\mu j^\mu = 0 \tag{8.77}$$

which, given that it involves 4-vectors, will have the same form in every frame, as desired. This reveals $\boldsymbol{J}$ in any frame to be the 3-vector that is needed to conserve charge density which, by definition, is the current density!

## 8.9 The Covariant Maxwell Equations

We are now ready to state the so-called *Lorentz covariant* form of the Maxwell equations (or just *covariant Maxwell equations* for short). Here the word "covariant" simply means "transforms nicely under Lorentz transformations", meaning that the equations are written in terms of 4-vectors and tensors. There are two equations, and the first is

$$\partial^\mu F_{\mu\nu} = \mu_0 j_\nu. \tag{8.78}$$

If you have not seen this before, it will look utterly mysterious. But the fact that this *could* correspond to some of Maxwell's equations is not so daft: the left-hand side contains a derivative acting on the field strength tensor which, as we have seen in Eq. (8.74), contains the electric and magnetic fields. Furthermore, the right-hand side of Eq. (8.78) contains the charge and current densities that we know should be there in Maxwell's equations. It thus seems plausible that Eq. (8.78) might correspond to Eqs. (7.16) and (7.19). This is indeed the case, as we now show. Note that the left-hand side of Eq. (8.78) contains the derivative operator with an *upstairs* index rather than a downstairs one. Given that turning an upstairs index

[1]Note in Eq. (8.77) that the factor of $c$ in the zeroth component of the 4-current density neatly cancels with that in the zeroth component of the derivative operator.

into a downstairs ones flips the sign of the spacelike components of a 4-vector, it follows that we must have

$$\partial^\mu = \left(\frac{1}{c}\frac{\partial}{\partial t}, -\nabla\right). \tag{8.79}$$

Also, the right-hand side of Eq. (8.78) requires the current density with a downstairs index, which from Eq. (8.75) must be

$$j_\nu = (c\rho, -\boldsymbol{J}). \tag{8.80}$$

We can now view Eq. (8.78) as four equations, one for each value of the free index $\nu$. Let us take $\nu = 0$. The implicit sum over $\mu$ then implies

$$\partial^0 F_{00} + \sum_{i=1}^{3} \partial^i F_{i0} = \mu_0 j_0.$$

Reading off $F_{00}$ and $F_{i0}$ from Eq. (8.74), and $\partial^i$, $j_0$ from Eqs. (8.79) and (8.80), this becomes

$$\sum_{i=1}^{3}(-\nabla_i)\left(-\frac{E_i}{c}\right) = c\mu_0\rho,$$

which simplifies upon using Eq. (7.27) to give Eq. (7.16), as required.

Now let us take $\nu = 1$ in Eq. (8.78), which gives

$$\begin{aligned}\partial^0 F_{01} + \sum_{i=1}^{3} \partial^i F_{i1} &= \frac{1}{c}\frac{\partial}{\partial t}\left(\frac{E_x}{c}\right) + \partial^2(B_z) + \partial^3(-B_y)\\ &= \frac{1}{c^2}\frac{\partial E_x}{\partial t} - \left(\frac{\partial B_z}{\partial y} - \frac{\partial B_y}{\partial z}\right)\\ &= \mu_0 j_x = -\mu_0 J_x.\end{aligned}$$

Rearranging and using Eq. (7.27), we obtain the $x$-component of Eq. (7.19). The $y$ and $z$ components can be obtained in a similar fashion from Eq. (8.78) by setting $\nu = 2$ and $\nu = 3$, respectively.

We see that Eq. (8.78) gives us two of the Maxwell equations, but what about the other two? These turn out to follow from the definition of the field strength tensor of Eq. (8.67), which in turn implies

$$\partial_\alpha F_{\mu\nu} + \partial_\nu F_{\alpha\mu} + \partial_\mu F_{\nu\alpha} = 0. \tag{8.81}$$

This is known as the *Bianchi identity*, and to see where it comes from, we can simply substitute Eq. (8.67) to obtain

$$\begin{aligned}\partial_\alpha F_{\mu\nu} + \partial_\nu F_{\alpha\mu} + \partial_\mu F_{\nu\alpha} &= \partial_\alpha\partial_\mu A_\nu - \partial_\alpha\partial_\nu A_\mu + \partial_\nu\partial_\alpha A_\mu - \partial_\nu\partial_\mu A_\alpha \\ &\quad + \partial_\mu\partial_\nu A_\alpha - \partial_\mu\partial_\alpha A_\nu.\end{aligned}$$

The order in which we carry out partial derivatives does not matter, and we see that the six terms on the right-hand side then combine in pairs that cancel out.

Now let us set $(\alpha, \mu, \nu) = (0, i, j)$ in Eq. (8.81), where $i, j \neq 0$:

$$\begin{aligned}\partial_0 F_{ij} + \partial_j F_{0i} + \partial_i F_{j0} &= \frac{1}{c}\frac{\partial}{\partial t}F_{ij} + \partial_j\left(\frac{E_i}{c}\right) + \partial_i\left(-\frac{E_j}{c}\right) \\ &= \frac{1}{c}\left[\frac{\partial}{\partial t}F_{ij} + \partial_j E_i - \partial_i E_j\right] = 0.\end{aligned}$$

Taking $(i, j)$ to be pairs of distinct values $(1, 2)$, $(2, 3)$ or $(1, 3)$, we then recover Eq. (7.18). Finally, let us set $(\alpha, \mu, \nu) = (1, 2, 3)$. We then get

$$-(\partial_1 F_{23} + \partial_2 F_{31} + \partial_3 F_{12}) = \frac{\partial B_x}{\partial x} + \frac{\partial B_y}{\partial y} + \frac{\partial B_z}{\partial z} = \nabla\cdot\boldsymbol{B} = 0,$$

which is Eq. (7.17). We have thus reproduced all of Maxwell's equations from Eqs. (8.78) and (8.81). Note that it makes sense that Eq. (8.81) reproduces Eqs. (7.17) and (7.18): the latter equations do not involve the charge or current density, which is also absent in Eq. (8.81). We can also now validate the assumption that we made above that Eq. (8.62) constitutes a 4-vector. Given that the right-hand side of Eq. (8.78) is a 4-vector, the left-hand side must also transform appropriately under Lorentz transformations. This implies that $F_{\mu\nu}$ is a bona fide tensor, which in turn implies that $A^\mu$ is a 4-vector!

Above, we saw that any equation written in terms of 4-vectors and tensors will have the same form in any frame. Both Eqs. (8.78) and (8.81) have this property, and thus are usually called the covariant Maxwell equations, as arbitrary Lorentz transformations do not change the *form* of the equations (although the explicit components of each 4-vector or tensor will change). If we instead use the 3-vector formalism of Eqs. (7.16)–(7.19), the form of the equations will be the same in all frames. But it is then unclear how the $\boldsymbol{E}$ and $\boldsymbol{B}$ fields in one frame are related to those in another. To see how this works explicitly, let us take the special case of frame $S'$

moving with speed $v$ along the positive $x$-axis with respect to a frame $S$, for which the appropriate Lorentz transformation is given in matrix form in Eq. (8.24). We know that the electric and magnetic field components in frame $S$ are given by components of the field strength tensor in this frame, as in Eq. (8.74). From Eq. (8.55), we know that the field strength tensor in the new frame $S'$ is given by

$$F'_{\mu\nu} = (\Lambda^{-1})^{\alpha}{}_{\mu}(\Lambda^{-1})^{\beta}{}_{\nu}F_{\alpha\beta}, \tag{8.82}$$

where the appropriate inverse Lorentz transformation can be found in Eq. (8.31). Equation (8.82) becomes easier to interpret if we rewrite it in matrix form. To do so, note that the upper and lower indices in each inverse Lorentz matrix label the rows and columns, respectively, according to the usual rules of matrix indices. In order to multiple matrices in the correct order, the row and column indices must be next to each other in index notation. For the first factor on the right-hand side, we may write

$$(\Lambda^{-1})^{\alpha}{}_{\mu} = [(\Lambda^{-1})^{\mathrm{T}}]_{\mu}{}^{\alpha},$$

where T denotes the transpose, which interchanges rows and columns, and thus the ordering of the row and column indices. Equation (8.82) can then be rewritten as

$$F'_{\mu\nu} = [(\Lambda^{-1})^{\mathrm{T}}]_{\mu}{}^{\alpha}F_{\alpha\beta}(\Lambda^{-1})^{\beta}{}_{\nu}. \tag{8.83}$$

Here we have reordered factors, which we are allowed to do given that each of these is an individual component of a matrix, and thus a number. Things are now such that all repeated indices (which are summed over due to the summation convention) occur next to each other. Thus, we can interpret Eq. (8.83) in matrix form as

$$\mathsf{F}' = (\Lambda^{-1})^{\mathrm{T}}\,\mathsf{F}\,\Lambda^{-1}. \tag{8.84}$$

Substituting in the explicit forms of Eqs. (8.74) and (8.31), this translates as

$$\begin{pmatrix} 0 & E'_x/c & E'_y/c & E'_z/c \\ -E'_x/c & 0 & -B'_z & B'_y \\ -E'_y/c & B'_z & 0 & -B'_x \\ -E'_z/c & -B'_y & B'_x & 0 \end{pmatrix}$$

$$= \begin{pmatrix} \gamma & +\gamma\beta & 0 & 0 \\ +\gamma\beta & \gamma & 0 & 0 \\ 0 & 0 & 1 & 0 \\ 0 & 0 & 0 & 1 \end{pmatrix} \begin{pmatrix} 0 & E_x/c & E_y/c & E_z/c \\ -E_x/c & 0 & -B_z & B_y \\ -E_y/c & B_z & 0 & -B_x \\ -E_z/c & -B_y & B_x & 0 \end{pmatrix}$$

$$\times \begin{pmatrix} \gamma & +\gamma\beta & 0 & 0 \\ +\gamma\beta & \gamma & 0 & 0 \\ 0 & 0 & 1 & 0 \\ 0 & 0 & 0 & 1 \end{pmatrix}, \tag{8.85}$$

where $\boldsymbol{E}'$ and $\boldsymbol{B}'$ are the electric and magnetic field vectors in $S'$. Equation (8.85) directly expresses these in terms of the corresponding quantities $(\boldsymbol{E}, \boldsymbol{B})$ in $S$. Multiplying out the matrices on the right-hand side is straightforward, albeit tedious! However, we eventually find

$$E'_x = E_x, \quad E'_y = \gamma(E_y - vB_z), \quad E'_z = \gamma(E_z + vB_y) \tag{8.86}$$

and

$$B'_x = B_x, \quad B'_y = \gamma\left(B_y + \frac{v}{c^2}E_z\right), \quad B'_z = \gamma\left(B_z - \frac{v}{c^2}E_y\right). \tag{8.87}$$

These are the transformation equations that relate electromagnetic fields in one frame, with those in another. Furthermore, these expressions are perhaps more general than they first appear, given that we are always free to rotate our coordinate system so that the boost between frames is in the $x$ direction. In order to interpret these equations, note that they make precise the physical intuition that we developed in earlier chapters: if static charges create only electric fields and moving charges also create magnetic fields, it must be the case that electric and magnetic fields mix up with each other under Lorentz transformations, which turn static charges into moving ones! The fact that this mixing is described by Eqs. (8.86) and (8.87) is not obvious, and the derivation of these equations is greatly helped by the covariant formalism of Eqs. (8.78) and (8.81).

So far, we have seen how to reformulate the Maxwell equations in a manifestly covariant language. This does not completely describe electromagnetism, however, as we must also say how charged particles respond to electromagnetic fields. This is the content of the Lorentz force law of Eq. (5.2), and we may also see whether this can be obtained using a relativistic approach in terms of 4-vectors and tensors. First, however, we may note that forces in relativity theory are different to those in Newtonian mechanics, given that the definitions of energy and momentum also have to change. Recall that force in Newtonian mechanics can be expressed as the rate of change of momentum. In relativistic mechanics, we may instead consider the rate of change of the 4-momentum of Eq. (8.50):

$$F^\mu = \frac{dp^\mu}{d\tau}, \tag{8.88}$$

where, in order for this to be a 4-vector, we have differentiated with respect to the proper time $\tau$, as we did in forming the 4-velocity $u^\mu$ in Eq. (8.46). Now consider a charged particle with 4-velocity $u^\mu$, mass $m$ and charge $q$. Noting that the Lorentz force law relates the force on a charged particle to the latter's velocity, we must find a relativistic relation between the rate of change of 4-momentum of Eq. (8.88), the electric and magnetic fields that may be present (represented by the field strength tensor $F_{\mu\nu}$), and the 4-velocity. Furthermore, the "force" should be linear in the field strength tensor, and the 4-velocity. We are thus led to the following equation:

$$\frac{dp_\mu}{d\tau} = qF_{\mu\nu}u^\nu, \tag{8.89}$$

as a consistent way of combining the quantities we need. To see how the Lorentz force law comes out of this, let us write Eq. (8.89) in matrix form:

$$\begin{aligned}\frac{d}{d\tau}\begin{pmatrix} E/c \\ -p_x \\ -p_y \\ -p_y \end{pmatrix} &= q\begin{pmatrix} 0 & E_x/c & E_y/c & E_z/c \\ -E_x/c & 0 & -B_z & B_y \\ -E_y/c & B_z & 0 & -B_x \\ -E_z/c & -B_y & B_x & 0 \end{pmatrix}\begin{pmatrix} \gamma c \\ \gamma v_x \\ \gamma v_y \\ \gamma v_z \end{pmatrix} \\ &= q\gamma\begin{pmatrix} (v_xE_x + v_yE_y + v_zE_z)/c \\ -E_x - v_yB_z + v_zB_y \\ -E_y - v_zB_x + v_xB_z \\ -E_z - v_xB_y + v_yB_x \end{pmatrix},\end{aligned} \tag{8.90}$$

where we have used Eqs. (8.48), (8.51) and (8.74), and remembered that lowering the index of the 4-momentum reverses the sign of the space-like components. We may write the second line of Eq. (8.90) in a more compact notation by recognising the form of the dot and cross (vector) products in Cartesian coordinates:

$$\frac{d}{d\tau}\begin{pmatrix} E/c \\ -\boldsymbol{p} \end{pmatrix} = \gamma q\begin{pmatrix} \boldsymbol{v}\cdot\boldsymbol{E}/c \\ -\boldsymbol{E} - \boldsymbol{v}\times\boldsymbol{B} \end{pmatrix}. \tag{8.91}$$

Finally, using Eq. (8.47) on the left-hand side and rearranging, we find the two separate conditions

$$\frac{dE}{dt} = q\boldsymbol{E}\cdot\boldsymbol{v}, \quad \frac{d\boldsymbol{p}}{dt} = q(\boldsymbol{E}+\boldsymbol{v}\times\boldsymbol{B}). \tag{8.92}$$

The second of these is the Lorentz force law of Eq. (5.2), with one slight caveat: the 3-momentum appearing in Eq. (8.92) is the *relativistic* 3-momentum $\boldsymbol{p} = \gamma m\boldsymbol{v}$, that reduces to its non-relativistic counterpart for sufficiently low speeds relative to the speed of light. Thus, interestingly, we see that the Lorentz force law is in fact more general than Eq. (5.2), and applies for full relativistic mechanics, provided we take the correct momentum. To interpret the first equation in Eq. (8.92), let us again consider the non-relativistic limit, such that the energy reduces to the usual Newtonian energy. In a small time $dt$, the work done on our charged particle will be given by

$$dW = \boldsymbol{F}\cdot d\boldsymbol{x} = \boldsymbol{F}\cdot\boldsymbol{v}dt, \tag{8.93}$$

where $\boldsymbol{F}$ is the total electromagnetic force acting on it, and $d\boldsymbol{x} = \boldsymbol{v}dt$ is the distance it moves in time $dt$, which is in turn related to the particle's velocity $\boldsymbol{v}$. The change in energy of the particle must be equal to the work done on it, and we thus find

$$dW = dE \quad \Rightarrow \quad \frac{dE}{dt} = \boldsymbol{F}\cdot\boldsymbol{v}. \tag{8.94}$$

Substituting the expression for the Lorentz force, we find

$$\frac{dE}{dt} = q(\boldsymbol{E}+\boldsymbol{v}\times\boldsymbol{B})\cdot\boldsymbol{v} = q\boldsymbol{E}\cdot\boldsymbol{v}, \tag{8.95}$$

where the magnetic field term vanishes given that $\boldsymbol{v}\times\boldsymbol{B}$ is orthogonal to $\boldsymbol{v}$. This amounts to the first equation of Eq. (8.92), which we can thus interpret as conservation of energy. Outside the non-relativistic regime, this equation remains true, but where $E$ is the full relativistic energy.

We have now succeeded in showing that the complete theory of electromagnetism can be written in terms of quantities (4-vectors and tensors) that transform naturally under the Lorentz transformations of special relativity. Historically, the equations of electromagnetism were written down before special relativity had been discovered, due to the fact that the former theory motivated the formulation of the latter. But it is interesting to ponder whether electromagnetism would have been arrived at much sooner had special relativity been known about beforehand. The requirements of SR seem to be highly constraining: electricity and magnetism must be unified from the outset, and the number of equations involved in the covariant formulation of the theory are less than the number of Maxwell equations in the non-covariant approach. Even then, the fact that there are so few equations governing such a wide array of physical phenomena suggests that there may be some more powerful principle underlying the structure of the theory. Indeed there is, as we explore in the following chapter.

## Exercises

(1) Consider a length $L_0$ and time difference $\tau_0$ measured in the frame $S'$ shown in Figure 8.1. Using the Lorentz transformations of Eq. (8.2), show that in the frame $S$, the length $L$ and time difference $\tau$ are contracted and dilated respectively, according to the formulae

$$L = \frac{L_0}{\gamma}, \quad \tau = \gamma\tau_0.$$

(2) Prove the vector calculus identity in Eq. (8.60).

(3) Show that raising one or two indices of the field strength tensor $F_{\mu\nu}$ leads to the expressions

$$F^{\mu}{}_{\nu} = \begin{pmatrix} 0 & E_x/c & E_y/c & E_z/c \\ E_x/c & 0 & B_z & -B_y \\ E_y/c & -B_z & 0 & B_x \\ E_z/c & B_y & -B_x & 0 \end{pmatrix},$$

$$F^{\mu\nu} = \begin{pmatrix} 0 & E_x/c & E_y/c & E_z/c \\ E_x/c & 0 & -B_z & B_y \\ E_y/c & B_z & 0 & -B_x \\ E_z/c & -B_y & B_x & 0 \end{pmatrix}.$$

(4) Consider a positive point charge in a frame $S'$ at which it is at rest. Find the magnetic field $\boldsymbol{B}$ in a frame $S$ defined as in Figure 8.1, in terms of electromagnetic field components in $S'$. Which direction does $\boldsymbol{B}$ point in, and why?

# Chapter 9

# Maxwell's Equations from Symmetry

The covariant forms of the Maxwell equations that we saw in Chapter 8 are very compelling. For example, they make precise the notion that electricity and magnetism must be combined into a single theory, given that the $\boldsymbol{E}$ and $\boldsymbol{B}$ fields are unified into a single object (the field strength tensor). However, the route to Eqs. (8.78) and (8.81) may have looked somewhat arbitrary. Why do we define the field strength tensor as in Eq. (8.67), and not through some other choice? Is this merely a rewriting of the Maxwell equations that makes relativity manifest, or is there some deeper principle underlying the theory, that makes Eqs. (8.78) and (8.81) *inevitable?* There is indeed such an underlying principle, namely an abstract mathematical symmetry called (*local*) *gauge invariance* that, if imposed on the theory describing matter in our universe, unavoidably leads to the existence of electromagnetism. Furthermore, this powerful idea can be generalised, such that it led to the development of the theory describing three of the fundamental forces in nature, namely the Standard Model of Particle Physics. This is a type of theory known as a *quantum field theory*, a full understanding of which would take us way beyond the scope of this book! Instead, we hope in this chapter to sketch the main ideas behind gauge invariance and how it constrains our theories of forces, in terms that can be appreciated at a lower level, provided one takes certain statements on trust. First, we need to understand how the matter in our universe is described using quantum field theory.

## 9.1 The Need for Quantum Field Theory

All of the known matter in our universe is composed of a relatively small number of building blocks. Inside the atom, for example, are electrons

Table 9.1 The range of physical theories applying to different physical situations.

| | Speed of object | |
|---|---|---|
| Size of object | Slow | Fast |
| Big | Newtonian Mechanics | Special Relativity |
| Small | Quantum Mechanics | Quantum Field Theory |

*Note:* All theories in the upper and left-hand panels emerge as limits of quantum field theory.

orbiting a central nucleus, itself composed of protons and neutrons, collectively known as *nucleons*. The latter are composite, with other particles living inside them. These are called *quarks* which, together with the electron, constitute *fundamental particles*, with no internal structure to the best of our knowledge. Also existing in nature are heavier partners of the electron, called the muon and the tauon. Finally, there are very light particles called *neutrinos*, one for each electron-like particle, and which were first discovered in the radioactive decays of certain nuclei. The complete list of fundamental matter particles then comprises the quarks (of which there are six different types in total), and the leptons (the electron, muon, tauon and their associated neutrinos).

As you may already know, one cannot correctly describe physics at subatomic scales without replacing our everyday laws of motion with *quantum mechanics* (*QM*). These laws look wildly different to, e.g. Newton's laws of motion, but must somehow reduce to the latter for everyday distance scales. Likewise, we know that we must include the effects of Special Relativity (SR) if objects are moving fast enough, relative to the speed of light. Again, however, it must be true that SR reduces to Newtonian mechanics for slow enough speeds. This situation is depicted in Table 9.1, and to complete the table, there must be a theory that allows us to include *both* SR and QM. This is precisely what quantum field theory does. Although there are other types of theory that also do this (e.g. string theory, which you may have heard of), we know on very general grounds that *any* such exotic theory must look like a quantum field theory at sufficiently low energies. To understand the basic idea of quantum field theory, let us temporarily leave the subject of matter particles, and consider the electromagnetic field. Historically, this was the context in which quantum theory was first discovered, at the turn of the twentieth century. In trying to explain the spectrum of wavelengths of electromagnetic radiation emitted by hot objects, Planck assumed that electromagnetic waves of a given frequency $\nu$ could not carry arbitrary energy, but that this energy must instead be "quantised" in basic units

$$E = h\nu, \tag{9.1}$$

where $h$ became known as *Planck's constant*, and has measured value (in SI units) $h = 6.63 \times 10^{-34}$ Js. Around the same time, Einstein found that the same assumption could explain the so-called *photoelectric effect*, whereby electrons are emitted from metals that have been illuminated by electromagnetic radiation. It took many decades to arrive at a consistent theory that includes the quantisation of radiation, but our modern understanding can be summarised quite simply. Light, as we know from Chapter 7, is an electromagnetic wave, which is a particular solution of the Maxwell equations for the electromagnetic field. In the quantum theory, light waves do not have continuous energy, but consist of single *photons*, carrying the discrete energy mentioned above. A given beam of electromagnetic radiation (of specific wavelength) will consist of multiple photons, and thus its energy will be quantised in multiples of the basic photon energy. In some sense, we can think of the photon as a "particle" of light and thus, in turn, of the electromagnetic field. However, what we mean by "particle" in this context is very different to the particles of Newtonian mechanics. The phrase "quantum field theory" now also makes sense: it is the theory that describes how (in this case) the electromagnetic field is not in fact continuous, but actually has a discrete (quantum) character, in turn giving rise to a particle-like interpretation. To summarise, the basic ideas are: (i) electromagnetism is described by a field filling all space; (ii) the equations for this field (Maxwell's equations) have wave-like solutions; (iii) these wave-like solutions cannot have continuous energy in the quantum theory, but instead arrive in discrete lumps ("particles") called photons.

Remarkably, these ideas generalise to all of the matter and forces we see in the universe. Every type of matter we know about is associated with a field filling all spacetime: there is an electron field, there are quark fields, and so on. Each of these fields is described by some equations, which are of course not the same as the Maxwell equations in general (the different types of matter are clearly not the same as light!). However, all of these equations have wave-like solutions, such that in the quantum theory, these waves do not have continuous energy, but instead have particle-like quanta associated with them. All of the known matter and force particles arise in this way, such that everything we see around us is ultimately a consequence of oscillating fields! This sounds extremely strange — and perhaps a little humbling — to the uninitiated. However, there is also something very natural about quantum field theory, in that it posits a single type of object (the field), from which everything else (waves and/or particles) emerges.

## 9.2 Matter and the Dirac Equation

Above, we listed the various different types of matter particle — namely the quarks and leptons. All of these particles have a curious quantum property called *spin*, that we have already discussed when talking about how electrons in solids lead to magnetised materials, in Chapter 5. That is, the fundamental matter particles have what looks like an intrinsic angular momentum. Rather than being continuous, the "spin" angular momentum can only take discrete values, namely an integer or half-integer number times $\hbar$, where

$$\hbar \equiv \frac{h}{2\pi},$$

and $h$ is the same Planck constant that we encountered above. A particle with intrinsic angular momentum $s\hbar$ is referred to as a *spin-$s$* particle, where $s \in \{0, 1/2, 1, \ldots\}$. All of the fundamental matter particles happen to be spin-1/2, and we know from above that there must be a field associated with each such particle, with an associated field equation. For free spin-1/2 particles that are not interacting with anything, the relevant field equation is the *Dirac equation*, that was first written down (in a slightly different context) in the 1920s. For the purposes of this chapter, we shall simply quote the equation, without further justification. However, the interested reader will find a more thorough explanation of where the equation comes from in Appendix B.

We will consider a single matter particle — the electron — and write the corresponding field as $\Psi(x)$. Here $x$ is the 4-vector position, such that the field can depend on space and time in general, as can the electromagnetic field. It turns out that $\Psi(x)$ is not a simple scalar quantity, but is actually a 4-component object called a *spinor*. None of this matters for the following argument, but we explain why this has to be the case in Appendix B. The Dirac field equation for $\Psi$ is then as follows:

$$(i\gamma^\mu \partial_\mu - m)\Psi(x) = 0, \tag{9.2}$$

where $m$ is the mass of the electron, and $\gamma^\mu$ a 4-vector containing certain constants.[1] Note that, in quoting this equation, we have adhered to a conventional system of units that are widely used throughout particle physics. In so-called *natural units*, explicit factors of $\hbar$ and $c$ (also $\mu_0$ and $\epsilon_0$, which $c$ depends on) are ignored, given that these can in principle be reinstated using dimensional analysis. This greatly simplifies the equations we will see throughout this and the following chapter. To convert any equation from elsewhere in the book into natural units, one can simply make the replacements

$$\hbar, c, \mu_0, \epsilon_0 \rightarrow 1. \tag{9.3}$$

We can see a factor of the complex $i$ in Eq. (9.2), and indeed the field $\Psi(x)$ turns out to be complex in general. This is not a problem, as it always works out that any physically measurable quantities are real numbers. However, the complex nature of $\Psi$ means that it carries a phase in general, which may be different at different points in spacetime. This leads to an interesting symmetry, which we discuss in the following section.

## 9.3 Global Gauge Invariance

The phase of a complex number can take values from 0 to $2\pi$, such that we may represent the phase of our electron field at each point in spacetime by an arrow on a circle. In Figure 9.1, we show two different points in spacetime, and represent the phase of the electron field at each point in this way (black arrows). Of course, where we choose to set the zero of phase is up to us. Choosing a different point on the circle to represent zero phase is the same thing as keeping the circle fixed, but rotating the arrows. Thus, we are clearly free to rotate the arrows at every spacetime point by the *same* amount, and to regard this as a physically indistinguishable electron field. This is shown by the dashed arrows in Figure 9.1, and we can make this idea precise as follows. Rotating a complex number by an angle $\alpha$ amounts to multiplying it by a complex number $e^{i\alpha}$. Thus, changing the phase of the electron field at all spacetime points simultaneously amounts

[1]Given that $\Psi$ is actually a 4-component object, $\gamma^\mu$ must be a vector of $4 \times 4$ matrices acting on $\Psi(x)$, and there is also an implicit $4 \times 4$ identity matrix in the mass term. Again this is unimportant for the arguments of this chapter, and more details are given in the appendix.

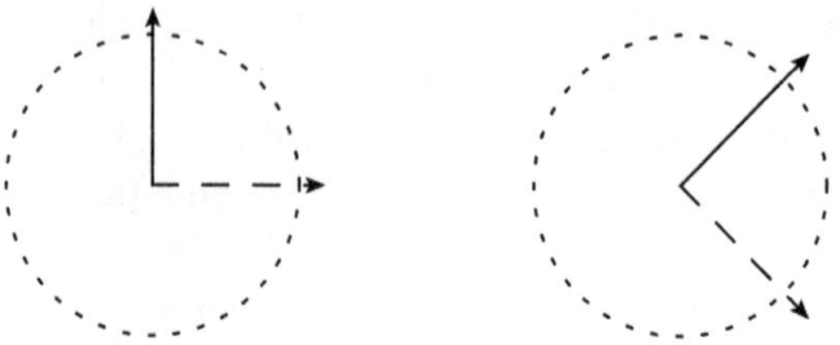

Fig. 9.1 The phase of the electron field at two different points in spacetime (black arrows). Also shown is the effect of a global gauge transformation (dashed arrows), which rotates the arrows by the same fixed amount at different points.

to the transformation

$$\Psi(x) \to e^{i\alpha}\Psi(x), \tag{9.4}$$

for all spacetime points $x$, where $\alpha$ is constant. Indeed, substituting this into Eq. (9.2) yields

$$(i\gamma^\mu \partial_\mu - m)e^{i\alpha}\Psi(x) = e^{i\alpha}(i\gamma^\mu \partial_\mu - m)\Psi(x) = 0,$$

so that our field remains a solution of the Dirac equation. A choice of how to fix the zero of phase is referred to as a choice of *gauge*, such that Eq. (9.4) is known as a *global gauge transformation.* The word "global" in this context refers to the fact that the phase is transformed by the same amount at all points in spacetime. The fact that this does not change the physics of a particular solution for the electron field can be fancily described by saying that "the theory of the electron is invariant under global gauge transformations".

## 9.4 Local Gauge Invariance

So far so good. But the actual theory of electrons interacting via electromagnetism turns out to have a much more powerful symmetry. Let us reconsider the transformations of Eq. (9.4), but now make the phase change *different* at different points in spacetime. This is called a *local gauge transformation* to distinguish it from the global case above, and it mathematically amounts to letting the transformation parameter $\alpha$ depend on the spacetime position: $\alpha \equiv \alpha(x)$. It is straightforward to see that Eq. (9.2) is not invariant under this transformation. Substituting the local form of

Eq. (9.4) into Eq. (9.2) yields

$$(i\gamma^\mu \partial_\mu - m)e^{i\alpha(x)}\Psi(x) = e^{i\alpha}\left[i(\gamma^\mu\partial_\mu - m) - (\partial_\mu \alpha)\right]\Psi.$$

We now have an extra term involving $\alpha$, whose origin is that the derivative $\partial_\mu$ acts on the local phase parameter $\alpha(x)$, as well as the electron field $\Psi$. This means that the transformed equation no longer has the same form as the original equation, so that the theory is not invariant under such transformations. It is not clear at this point *why* we might want to make the theory locally gauge invariant, but let us for now note that we can manage this by finding a new operator $D_\mu$, such that

$$D_\mu \Psi \rightarrow e^{i\alpha(x)} D_\mu \Psi \tag{9.5}$$

under a local gauge transformation $\Psi \rightarrow e^{i\alpha}\Psi$. If we then replace the derivative $\partial_\mu$ with $D_\mu$, we will have succeeded in constructing a locally gauge-invariant Dirac equation. To see this, note that the modified Dirac equation transforms under a gauge transformation according to

$$(i\gamma^\mu D_\mu - m)\Psi \rightarrow e^{i\alpha}(i\gamma^\mu D_\mu - m\Psi) = 0,$$

and the overall factor of $e^{i\alpha}$ can be cancelled, leading to the same equation as before. The operator $D_\mu$ is called the *covariant derivative*, and we can find it by making the following guess:

$$D_\mu = \partial_\mu + ieA_\mu(x), \tag{9.6}$$

where we have included the complex factor $i$, and the constant $e$, to agree with existing conventions. In the second term on the right-hand side, we have included an arbitrary vector field $A_\mu(x)$, which is a natural guess to make given that we must modify the normal derivative with something that has a single spacetime index $\mu$, and which may in general depend upon spacetime position. We wish to verify that Eq. (9.6) can be made to verify Eq. (9.5) under a local gauge transformation of $\Psi$ and, in doing so, we can also include the possibility that $A_\mu$ itself changes. Not knowing what the transformation of $A_\mu$ might be, however, we can simply write

$$\Psi(x) \rightarrow e^{i\alpha(x)}\Psi(x) \quad \Rightarrow \quad A_\mu(x) \rightarrow A_\mu(x) + \delta A_\mu(x),$$

for some change $\delta A_\mu$, that can depend upon spacetime position in general. The complete transformation of Eq. (9.6) is then

$$\begin{aligned} D_\mu\Psi(x) &\to (\partial_\mu + ieA_\mu + ie\delta A_\mu)e^{i\alpha(x)}\Psi(x) \\ &= e^{i\alpha(x)}\left[\partial_\mu\Psi + ieA_\mu + i\partial_\mu\alpha(x) + ie\delta A_\mu\right]\Psi(x) \\ &= e^{i\alpha}\left[D_\mu + i\partial_\mu\alpha(x) + ie\delta A_\mu\right]\Psi(x). \end{aligned}$$

It follows that we can satisfy the requirement of Eq. (9.5) provided we stipulate that $A_\mu$ change under gauge transformations as follows:

$$\delta A_\mu(x) = -\frac{1}{e}\partial_\mu\alpha(x). \tag{9.7}$$

This is eminently possible, and what we have thus shown is that — should we want to — we can construct a theory of the electron that is locally gauge invariant. However, this theory is not very satisfactory. In order that the electron field satisfy local gauge invariance, we have had to introduce the quantity $A_\mu(x)$ which, owing to its spacetime dependence, is clearly some sort of field itself. It is not particularly surprising that this happens: we have gained the ability to change the phase of the electron field arbitrarily at every spacetime point, which constitutes an infinite number of degrees of freedom. The price of achieving this is having to introduce an additional field, which itself contains an infinite number of degrees of freedom. When we do a local gauge transformation, the field $A_\mu$ is then able to "compensate" for each local change in phase. However, this field is not unique: we could find infinitely many quantities $A_\mu$ that satisfy the transformation law of Eq. (9.7). How can we make sense of such a theory?

Now comes the clever part. Let us demand that the electron theory is indeed locally gauge invariant, but let us then postulate that the field $A_\mu$, rather than being a convenient mathematical trick, is actually a real physical field that exists in nature.[2] Then we can construct a field equation for $A_\mu$, whose solution will tell us how $A_\mu$ behaves in any given situation. The field equation for $A_\mu$ should be gauge-invariant, and it seems natural that it should be second-order in derivatives, as are many other equations of motion. Thus, we want to take the object $A_\mu$, and construct a gauge-invariant object by combining it with the derivative operator $\partial_\mu$. The simplest choice $\partial_\mu A_\nu$ is not gauge-invariant, as under the transformation of

[2] We do not necessarily imply that $A_\mu$ is itself directly measurable. However, it must be related to things that are.

Eq. (9.7) one finds

$$\partial_\mu A_\nu(x) \to \partial_\mu A_\nu(x) - \frac{1}{e}\partial_\mu \partial_\nu \alpha(x).$$

However, if we instead try

$$F_{\mu\nu} = \partial_\mu A_\nu - \partial_\nu A_\mu,$$

then this is indeed gauge invariant: under Eq. (9.7) one finds

$$F_{\mu\nu} \to \partial_\mu A_\nu - \frac{1}{e}\partial_\mu \partial_\nu \alpha(x) - \partial_\nu A_\mu + \frac{1}{e}\partial_\nu \partial_\mu \alpha(x).$$

The two terms involving $\alpha(x)$ cancel each other, due to the fact that the order in which the derivatives are taken does not matter. We are then left with the original form for $F_{\mu\nu}$. If we now want a field equation that is linear in $A_\mu$, and second-order in derivatives, we are led to the general form

$$\partial^\mu F_{\mu\nu} = J_\nu, \tag{9.8}$$

which is precisely the form of the relativistic Maxwell equation we presented in Eq. (8.78). Furthermore, the Bianchi identity of Eq. (8.81) is automatically satisfied given the definition of $F_{\mu\nu}$. As we saw in the previous chapter, if we label the components of $F_{\mu\nu}$ in terms of 3-vector fields, Eqs. (8.78) and (8.81) give rise to the Maxwell equations of Eqs. (7.16)–(7.19). We are thus led to the remarkable conclusion that *demanding the theory of the electron satisfy local gauge invariance leads to the existence of electromagnetism!* The complete theory of the electromagnetic field summarised by the Maxwell equations took thousands of years to arrive at, not least due to the fact that it took humankind so long to realise that electricity and magnetism had to be combined into a single theory. In hindsight, much of this effort seems wasted, given that the complete equations describing the theory can be obtained from a single symmetry principle! Furthermore, gauge invariance provides some sort of explanation for *why* electromagnetism is the way it is, including why charge is conserved.[3]

For completeness' sake, we should point out that the choice of the gauge invariant quantity of Eq. (8.67) is not unique. We could have added an extra term to Eq. (8.67), which we shall not bother writing explicitly here, but which would give rise to the existence of magnetic monopoles. Thus,

[3]The more mathematical reader may have heard of *Noether's theorem*, which relates symmetries obeyed by a given theory to conserved quantities. The conserved quantity relating to gauge invariance turns out to be electromagnetic charge.

the apparent absence of magnetic monopoles in nature remains a mystery. Furthermore, the cynical reader may remark that gauge invariance is not really an explanation of where electromagnetism comes from: we now have to explain why local gauge invariance should be a property of the universe we live in. One justification for its existence is that, in a full quantum theory, the masses of particles get corrections from quantum effects. However, we know that the photon has to be precisely massless, and it turns out that gauge invariance is crucial in ensuring that this remains true. Also, naïve attempts to quantise the theory of electromagnetism lead to unphysical degrees of freedom that make the theory mathematically inconsistent. Gauge invariance removes these degree of freedom, so that only the physical polarisation states of the photon contribute.

From a practical point of view, gauge invariance means that there is a huge redundancy in the theory: there are an infinite number of choices for $A_\mu$ that give the same field strength $F_{\mu\nu}$, and hence the electric and magnetic fields $\boldsymbol{E}$ and $\boldsymbol{B}$. To see what this looks like in terms of the non-relativistic formalism of electromagnetism, note that Eqs. (9.7) and (8.63) imply that the electrostatic and magnetic vector potentials transform as follows:

$$V \to V - \frac{1}{e}\frac{\partial \alpha}{\partial t}, \qquad \boldsymbol{A} \to \boldsymbol{A} + \frac{1}{e}\nabla\alpha. \tag{9.9}$$

One may verify that this preserves the electric and magnetic fields defined by Eqs. (8.59) and (8.58). Choosing a particular form for $A_\mu$ is known as *fixing a gauge*, and often this is done by imposing an auxiliary condition on the field $A_\mu$. Examples include the *Coulomb gauge* $\nabla \cdot \boldsymbol{A} = 0$, and the *Lorenz gauge* $\partial_\mu A^\mu = 0$. Once a gauge has been fixed, one must use this throughout all calculations. Any physically observable quantity must be independent of the choice of gauge, which can often be used as a check on calculations, particularly in the full quantum field theory!

We did not yet discuss the 4-vector on the right-hand side of Eq. (9.8). Comparison with Eq. (8.78) shows that this will be interpreted as the current in our theory, and it is not arbitrary. Upon replacing the derivative operator in the Dirac equation with the covariant derivative of Eq. (9.6), the field equation for the electron becomes

$$(i\gamma^\mu \partial_\mu - m)\Psi(x) - e\gamma^\mu A_\mu(x)\Psi(x) = 0, \tag{9.10}$$

showing that the electron field is coupled to the electromagnetic field. We can thus interpret the constant $e$ as the charge of the electron, and consistency of the theory means that we must be able to find an expression for the current $J^{\mu}$ (which must be real) in terms of the electron field and its complex conjugate. This is indeed the case, as can be found in a suitable textbook on quantum field theory.

## 9.5 Gauge Invariance and the Other Forces

Throughout this chapter, we have seen that demanding that matter fields obey an abstract mathematical symmetry called local gauge invariance leads to the presence of a force (electromagnetism). It is then natural to ponder whether or not similar arguments can be used to derive the other forces in nature. Indeed, the weak and strong nuclear forces that are present in the Standard Model can also be obtained using different kinds of local gauge invariance! Whilst the mathematics of these forces is a lot more complicated than the electromagnetic case, the ideas are conceptually similar, and so we discuss them here.

Let us first recall the geometric interpretation of the phase of the electron field, depicted in Figure 9.1. It could be represented as an arrow in an abstract space at each spacetime point, namely the circles drawn in the figure. Local gauge invariance meant being able to choose the position of the arrow arbitrarily at different points in spacetime. This circle is sometimes referred to as an *internal space* (of the electron field), to distinguish it from the actual spacetime that the electron field lives in. More generally, we might consider a different type of arrow, in a more complicated internal space, and then construct a local gauge symmetry based on this. Let us consider the explicit example of the strong interaction, which is felt by quarks. It turns out that quarks have a type of charge called *colour*, which can take one of three values, conventionally called red ($r$), green ($g$) and blue ($b$). These names are merely labels, and colour charge has nothing to do with electromagnetic charge. The latter is associated with electromagnetism, whereas colour charge is associated with the strong force.

Quarks are spin-1/2 fields, and are thus described by the Dirac equation. However, at each point in spacetime, the field will have some amount each of red, green and blue charge. If we want to, we can draw this as an abstract arrow in an internal "colour space", as shown in Figure 9.2. Given that the names $r$, $g$ and $b$ are just labels, we are clearly free to relabel them, or mix them up, and thus we can rotate the coordinate axes in the colour space at whim. This is the same thing as keeping the axes fixed, but rotating the arrow, provided we rotate the colour arrows similarly at all points in

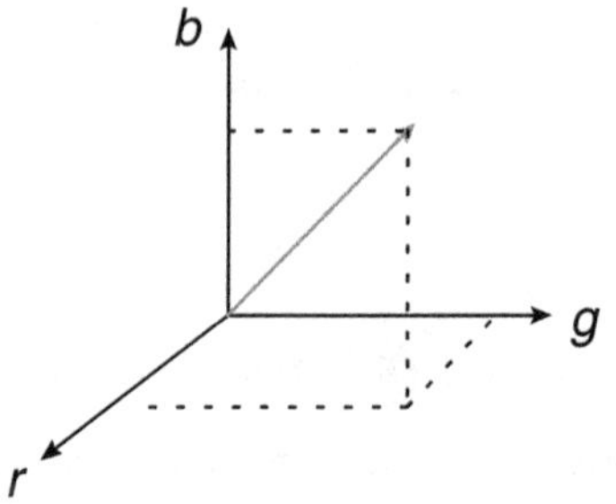


Fig. 9.2 At each spacetime point, the quark field has some amount of red, green and blue colour charge, which can be drawn as an arrow in an abstract internal *colour space.*

spacetime. Thus, the theory for the quark field must be invariant under global gauge transformations, which in this case consist of 3D rotations of the colour arrow that are the same at each spacetime point. This turns out to be a complex space, so that these are complex, rather than real, rotations. It is now hopefully easy to guess what the principle of local gauge invariance is in this case! It says that we are free to choose the direction of the colour arrow independently at each spacetime point. This in turn leads to the existence of a new field, which corresponds to the *gluon* that carries the strong interaction. What makes the mathematics significantly more complicated than the electromagnetic case is that we are now forced to embrace the full machinery for dealing with multidimensional (complex) symmetries. This is a branch of mathematics known as (Lie) group theory, and we refer the reader to textbooks on quantum field theory for a full discussion. However, it is perhaps useful to quote some of the terminology that you may see elsewhere. A feature of the electromagnetic gauge transformations discussed above is that the order in which they are performed is irrelevant: if we rotate an arrow twice around the same circle, we can carry out the rotations in either order. This is not true for rotations in three real dimensions, let alone three complex ones! Sets of symmetry transformations where the order does not matter are called *abelian groups.* By contrast, transformations such as those in the theory of the strong force are called *non-abelian groups.* Thus, you will often see electromagnetism and the strong force referred to as examples of an *abelian* and *non-abelian* gauge theory, respectively.

Things are yet more complicated for the weak interaction, which actually mixes with electromagnetism to make a single *electroweak theory* at high energies. The gauge theory describing this has an abelian part, and a non-abelian part, and the relevant local gauge symmetry gets broken at low energies, which turns out to be relevant for generating the masses of the so-called $W$ and $Z$ bosons that carry the weak force. An extra particle called the Higgs boson is needed to break the symmetry, and was relatively recently discovered (in 2012). Again, we refer the reader to modern quantum field theory textbooks for a full treatise.

In this chapter, we have reviewed the concept of local gauge invariance, and argued that is a powerful principle in "deriving" the forces that we see in nature. However, we are then forced to consider why the particular symmetries we see in nature are there, rather than other ones. Furthermore, why should these symmetries be there in the first place? Some theories of modern physics (e.g. string theory) have tried to derive the existence of gauge symmetries from more fundamental underlying principles. Other work has tried to reformulate quantum field theory so that gauge invariance is no longer needed from the outset. Curiously, we have so far completely ignored the fourth fundamental force in nature: gravity. We remedy this in the following chapter.

## Exercises

(1) Verify the gauge transformations of the electrostatic and magnetic vector potentials, given in Eq. (9.9).

(2) Consider imposing the Lorenz gauge condition

$$\partial_\mu A^\mu = 0.$$

(a) Show that the Maxwell equation of Eq. (8.78) reduces, in the vacuum case, to

$$\partial_\mu \partial^\mu A_\nu = 0.$$

(b) Show that this implies (in natural units)

$$\nabla^2 A_\nu = \frac{\partial^2 A_\nu}{\partial t^2}.$$

Interpret this result.

# Chapter 10

# The Double Copy: From Electromagnetism to Gravity

Throughout this book, we have reviewed the theory of electromagnetism, and explained how it is crucial to an understanding of the other forces of nature. In particular, it provided the first historical example of a gauge theory, and the principle of local gauge invariance then directly guided the formulation of quantum field theories of the weak and strong nuclear forces. The combination of these, together with a characterisation of all the fundamental matter particles in nature, constitutes the Standard Model of Particle Physics which, at the time of writing, has withstood every experimental test at particle accelerators.

So far, however, we have said comparitively little about gravity, despite the fact that this is one of the first forces we learn about as children. It causes objects to be held on the Earth's surface, but also controls the dynamics of our solar system. Further afield, gravity controls the workings of galaxies, clusters of galaxies and even the entire universe itself! At first glance, it may not seem that electromagnetism has much to say about gravity. Nor do particle collider experiments, given that the mass of subatomic particles is such that gravity is largely irrelevant. However, we know that our current best theory of gravity — Einstein's theory of General Relativity — is incomplete. It breaks down at extreme locations in spacetime, such as at the centre of black holes, or the origin of our universe itself (the "Big Bang"). It is widely believed that a quantum theory of gravity is needed in order to describe these situations, and to resolve the mystery of where the universe came from in the first place. But attempts to formulate a quantum field theory for gravity do not give satisfactory results. Alternative theories — such as string theory — are indeed able to unify gravity and (non-)abelian gauge theories into a single (quantum!) theoretical framework. However, despite decades of research on these theories, simple hopes of connecting them to concrete gravitational questions testable in current experiments have remained unfulfilled. That does not mean that such theories are of no use, as we will see below.

Even if quantum gravity turns out not to be a field theory, it must look like a field theory at sufficiently low energies, which may still be very high compared with our current experimental energy scales. It is also true that there are many open questions even in non-quantum ("classical") gravity, which would greatly benefit from new and efficient calculational techniques. Since the mid-2000s, new techniques for both classical and quantum gravity have originated from an unexpected quarter, namely the study of scattering particles in gauge theories such as those entering the Standard Model. There is now mounting evidence that our traditional methods for studying quantum field theories have been hiding a profound underlying structure, that makes such theories much simpler than previously thought. What's more, a series of truly amazing connections have been found between *different* field theories: for example, we now realise that much of what happens in gravity can be generated in a clever way by recycling gauge theory results! This makes electromagnetism a lot closer to active research in theoretical physics than it might seem.

The particular correspondence that we will review in this chapter is called the *double copy*, and was first discovered in the context of quantities called *scattering amplitudes* governing how particles interact in quantum field theory. Put simply, it states that formulae describing particle interactions in gravity (GR plus its generalisations) can be written in a form that looks almost identical to corresponding formulae in a non-abelian gauge theory, such as the theory of gluons in the Standard Model. Since then, the double copy has been extended to many different applications of classical gravity, including those relevant for astrophysics. The results are giving us new ways to calculate observables relevant for, e.g. gravitational wave experiments. But they also promise to completely change how we think about gravity, and how we unify it with the other interactions in nature. Let us first describe the context in which the double copy was first noticed.

## 10.1 Scattering Amplitudes in QFT

The scattering of particles has been of great historical importance in the formulation and testing of our theories of fundamental physics, including the Standard Model, and what may lie beyond it. A typical modern particle accelerator experiment consists of two beams of particles which are focused so that the constituent particles collide with each other. At the currently running Large Hadron Collider at CERN, for example, protons are collided, and the debris from each collision collected in large detectors. By analysing what happens, we can compare the predictions of the SM with those of other theories, in order to ascertain whether or not there is any evidence for physics beyond the SM.

In a quantum theory, it is not possible to say what will definitely happen in any collision, even if we were to know the incoming particle energies and momenta to infinite precision. Instead, for a given *initial state* consisting of two beam particles, there can be any number of *final states*, where each has a certain probability to occur. For each choice of final and initial states, the rules of QFT tell us how to calculate a number — the *scattering amplitude* $\mathcal{A}$ — that is related to the probability $P$ for the given final state to occur. The amplitude $\mathcal{A}$ is a complex number, and one has

$$P \sim |\mathcal{A}|^2,$$

such that the probability is manifestly positive (or zero), as required. In general, the amplitude is a complicated function of the 4-momenta of the incoming and outgoing particles, and also their other properties such as charges and polarisations. The fact that the amplitude is complex is perhaps not surprising: we know that particles in QFT arise as quanta of wave-like solutions. Waves carry a phase, so that we expect the amplitude itself to carry phase information, and thus to be complex in general.

In principle, scattering amplitudes can be calculated in QFT, although this can usually only be done approximately. Associated with each force in the Standard Model is a number called the *coupling constant*, which represents the strength of interaction between the matter and force particles. In electromagnetism, for example, this is the charge $e$ on the electron, as the electromagnetic charges of all other matter particles can be given in terms of this number. Here, we will concern ourselves with the theory of pure gluons. Unlike the theory of electromagnetism, in which photons are chargeless, gluons carry colour charge, so that they can interact with each other. This makes pure gluon theory highly non-trivial, and it is sometimes referred to as *Yang–Mills theory* in the literature, after the two people who discovered this theory, decades before it was realised it could be used to describe the gluon! Following convention, we will label the coupling constant in Yang–Mills theory by $g$, and we can then consider

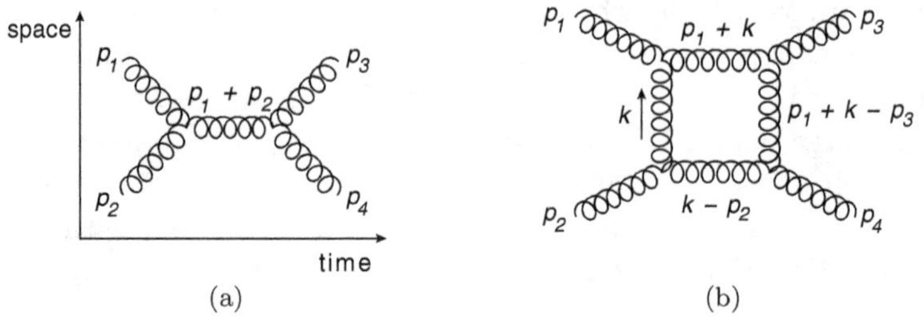


Fig. 10.1 Two example Feynman diagrams, showing the passage of various gluons and their interactions in space and time.

scattering processes involving other gluons. One way to represent these is to use *Feynman diagrams*, which are handy spacetime pictures representing the history of a given scattering event. For example, Figure 10.1(a) shows two gluons (shown as curly lines), that come together and annihilate to make a third gluon, which then travels some distance and decays. Figure 10.1(b) shows a more complicated process, in which gluons interact by exchanging multiple other gluons. In each case, we have external lines in the diagram, which represent real particles that emerge from the far past (e.g. as beams in an experiment), or travel to the far future (e.g. to a detector). There are also internal lines, representing intermediate particles in the scattering process. These particles are never seen directly, and are called *virtual* particles.

In Figures 10.1(a) and 10.1(b), we can label the 4-momentum of the particles at each stage, and the rules of QFT tell us that 4-momentum has to be conserved at each vertex. This leads to a crucial difference between the real particles (external lines) and virtual particles (internal lines). Whereas the former have 4-momentum satisfying the usual relativistic energy–momentum relation of Eq. (8.9), this cannot be true for the virtual particles. As an example, consider the intermediate particle in Figure 10.1(a), and let us parametrise the incoming 4-momenta as[1]

$$p_1 = (E_1, \boldsymbol{p}_1), \quad p_2 = (E_2, \boldsymbol{p}_2). \tag{10.1}$$

Then the squared 4-momentum of the internal line is given by

$$(p_1 + p_2)^2 = (E_1 + E_2, \boldsymbol{p}_1 + \boldsymbol{p}_2) \cdot (E_1 + E_2, \boldsymbol{p}_1 + \boldsymbol{p}_2)$$

[1]Throughout this chapter, we will again use the natural units discussed in the previous chapter, in which we ignore factors of $c$ and $\hbar$.

$$= (E_1 + E_2)^2 - (\boldsymbol{p}_1 + \boldsymbol{p}_2)^2$$
$$= E_1^2 - \boldsymbol{p}_1^2 + E_2^2 - \boldsymbol{p}_2^2 + 2(E_1 E_2 - \boldsymbol{p}_1 \boldsymbol{p}_2).$$

Like the photon, the gluon turns out to be massless, and thus one has

$$E_i^2 - \boldsymbol{p}_i^2 = 0 \quad \Rightarrow \quad E_i = |\boldsymbol{p}_i|$$

in natural units. This in turn implies

$$(p_1 + p_2)^2 = 2E_1 E_2 (1 - \cos\theta),$$

where $\theta$ is the angle between the 3-momenta of the two incoming particles. For two incoming beams with $\theta = \pi$, this is clearly not zero, and thus the virtual gluon *does not* obey the physical energy–momentum relation of Eq. (8.9). For this reason, real (virtual) particles are described as being *on (off) the mass shell*, respectively, or simply "on-shell" and "off-shell" for short.

There is another feature related to 4-momenta that it is worth drawing attention to. In Figure 10.1(a), we see that 4-momentum conservation fixes the internal line momentum purely in terms of the external 4-momenta. This is not true in Figure 10.1(b): we must supply a 4-momentum for one of the internal lines ($k$ in the figure), after which all other 4-momenta are fixed. If you look carefully, you will see that the issue is that Figure 10.1(b) has a loop in it, whereas Figure 10.1(a) does not. Diagrams with no loops are called *tree-level graphs*, and have the special property that all internal momenta are known from the external momenta. For loop-level graphs, every distinct loop requires that we introduce a *loop momentum*, which can then be used to work out all the other momenta.[2]

Feynman diagrams are much more than handy pictures that allow us to visualise how particle interactions work. There are precise rules in QFT that allow us to associate each diagram with a precise algebraic contribution to the scattering amplitude! A full discussion of these rules is clearly beyond the scope of this book, but some of them are straightforward to state if you are prepared to take them on trust. For Yang–Mills theory we have the

[2]Note that the procedure of conserving 4-momentum at each vertex of a Feynman diagram is very similar to the Kirchhoff law for currents at a junction, discussed in Chapter 4.

following:

(1) For a given initial and final state and number of loops, one must draw all possible Feynman diagrams containing the vertices allowed by the theory (in Yang–Mills theory, these are vertices connecting three or four gluons, although the latter can always be replaced with pairs of 3-gluon vertices if needed).
(2) Each vertex contributes a factor of the coupling constant $g$, as well as other factors depending on the 4-momenta of the gluons, and their colour charges.
(3) Each internal line $\alpha$ carries a factor of $1/p_\alpha^2$, where $p_\alpha$ is the 4-momentum of the internal line. Note that $p_\alpha^2 \neq 0$ given that internal lines are off-shell. There are also other factors in the numerator for each internal line, again potentially involving the 4-momenta of the gluons.
(4) External lines are associated with polarisation vectors for the gluons.
(5) We must integrate over all possible components of each loop 4-momentum $k_i$. This is due to the fact that, in a quantum theory, we must sum over all possible contributions to a given final state. Given that the loop momenta are never observed, we must sum over all possible loop momenta that give the same final state, which becomes an integral given that momenta are continuous.

This is a lot of work in general, and the result of all this effort is that a Yang–Mills scattering amplitude with $L$ loops and $m$ external particles (incoming or outgoing) assumes the form

$$\mathcal{A}_m^{(L)} \sim g^{m-2+2L} \sum_i \left( \prod_{l=1}^{L} \int d^d k_l \right) \frac{n_i c_i}{\prod_\alpha p_\alpha^2}, \tag{10.2}$$

where we have considered the general case of $d$ spacetime dimensions.

To understand this expression, note that the overall power of the coupling will indeed arise from dressing each vertex in a graph with a power of $g$. The sum is over distinct diagrams $i$ with the same initial and final state, and there are integrals over the loop momenta $\{k_l\}$ as required. From the above rules, there are denominator factors coming from each internal line in diagram $i$, as well as a numerator factor $c_i$ involving the colour charges of the gluons in each diagram. Finally, there is a *kinematic numerator factor* $n_i$, that will depend on the various 4-momenta in diagram $i$, as well as the polarisation states of the external gluons. This general form of the amplitude follows from the form of the Feynman rules, and it turns out that, for each diagram, the colour factor $c_i$ is extremely straightforward to find.

Thus, the task of calculating scattering amplitudes in quantum field theory be rephrased as needing to find the kinematic numerators $\{n_i\}$, before carrying out the integrals over the loop momenta. The kinematic numerators are not unique, but turn out to depend on the gauge used for the gluon field. More modern techniques try to dispense with traditional Feynman rules altogether, and work entirely in terms of gauge-invariant information from the outset.

From Eq. (10.2), we see that increasing the number of external lines or loops in a diagram involves a higher power of the coupling constant $g$. Feynman diagrams and rules thus amount to performing an expansion in $g$. This makes sense if the coupling constant is sufficiently small, which is indeed the case in many practical applications. We can then truncate our expression for the scattering amplitude at a given fixed order in $g$, and it is in this sense that most of our results for scattering amplitudes are approximate. This approach is known as *perturbation theory*, and something similar can also be carried out in (quantum) gravity. To see what this means in more detail, we must first review our current best description of gravity in terms of General Relativity (GR).

## 10.2 Gravity and General Relativity

In Chapter 8, we introduced the idea of *spacetime*, namely a 4D entity that unifies space and time into a single mathematical structure. This is the natural arena in which the laws of Special Relativity (SR) are formulated, given that Lorentz transformations mix up space and time. We have also constructed a theory of electromagnetism that is fully local in spacetime, the need for which arose in Chapter 3 when we discussed how Coulomb's law would otherwise imply action at a distance. The same objection could be raised about Newton's law of gravity, which says that the magnitude of the gravitational force between two objects with masses $m_1$ and $m_2$, and separated by a distance $r$, is given by

$$|\boldsymbol{F}| = \frac{G_N m_1 m_2}{r^2}, \tag{10.3}$$

where $G_N$ is *Newton's constant*, whose value is $6.67 \times 10^{-11}\,\mathrm{m^2\,kg}$ in SI units. According to this equation, if we move one of the masses, then the other mass appears to immediately "know" that the force on it has changed, despite the fact that $r$ may be arbitrarily large. Thus, Newtonian gravity is at odds with SR, in the same way that Coulomb's law proved to be. The

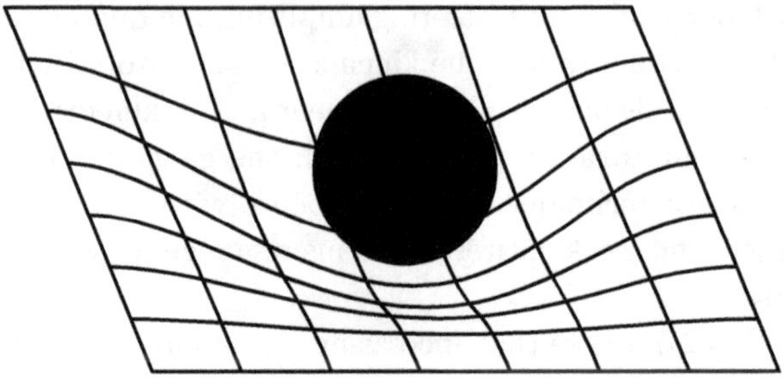

Fig. 10.2 In General Relativity, massive objects curve the space around them.

question then arises as to what the correct local theory of gravity should be, and the answer is much more complicated than you might think!

A local theory of gravity that is consistent with SR was first written down by Einstein in 1914, although it was based on a number of crucial insights by others. The key concept involved is that spacetime is no longer a passive observer to physics, as it is in special relativity. Instead, spacetime itself becomes a dynamical object, rather like a rubber sheet that can be stretched and warped. Put another way, conventional spacetime is *flat*, whereas the spacetime of GR is *curved* in general. It is precisely this curvature that appears as the force of gravity. Consider, for example, the heavy object shown in Figure 10.2, which deforms an otherwise flat space around it. A smaller test object in the vicinity of the heavy object would follow a curved path instead of a straight line, and be either deflected, or even pursue a closed orbit. It should be borne in mind that this analogy is imperfect: the curvature in GR is in 4D spacetime, rather than 3D space!

To understand GR further, we need to know how to describe curved spacetime mathematically. In fact, we have already set up most of the machinery for this in Chapter 8. In defining the dot product for 4-vector, we introduced the *metric tensor* $\eta_{\mu\nu}$. Imagine that we "zoom in" on a small region of spacetime, and consider an infinitesimal spacetime displacement $dx^\mu$. Then the spacetime "length" of this displacement, $ds$, satisfies

$$ds^2 = \eta_{\mu\nu} dx^\mu dx^\nu. \tag{10.4}$$

Thus, the "length" of a small displacement at some point in spacetime is directly governed by the metric tensor. In Minkowski space with Cartesian coordinates, the metric tensor is the same at all points in spacetime. However, if we take a warped space such as that shown in Figure 10.2, the lengths at different points in the spacetime may be different in general.

In other words, *the metric tensor should depend upon spacetime position.* Thus, to describe a curved space, we can replace the metric tensor appearing in Eq. (10.4) to give

$$ds^2 = g_{\mu\nu}(x)dx^\mu dx^\nu. \tag{10.5}$$

The quantity on the left is called the *line element*, and represents the proper length of a small displacement in a given coordinate system, at some point $x$ in spacetime. On the right, we have a general metric tensor $g_{\mu\nu}(x)$, whose dependence on the spacetime coordinates tells us that our spacetime is potentially warped, i.e. looks different at different spacetime points. Note that a metric tensor that depends upon position does not necessarily guarantee that a given space is curved. For example, the metric tensor for flat spacetime depends on the spatial coordinates if we use spherical polars, but this still represents a flat space. The branch of mathematics known as *differential geometry* describes how to talk about curved spaces in general, and it is this mathematics that underlies the theory of GR.

Equation (10.5) is written in a particular coordinate system, but the quantity on the left (the squared "length" of a 4-vector) does not depend on the coordinates. We are thus free to transform to a different coordinate system $x'^\mu$ on the right-hand side. Upon using the chain rule

$$dx^\mu = \frac{\partial x^\mu}{\partial x'^\rho} dx'^\rho, \tag{10.6}$$

one may rewrite Eq. (10.5) as

$$ds^2 = \left[ g_{\mu\nu}(x) \frac{\partial x^\mu}{\partial x'^\rho} \frac{\partial x^\nu}{\partial x'^\sigma} \right] dx'^\rho dx'^\sigma \equiv g'_{\mu\nu} dx'^\rho dx'^\sigma, \tag{10.7}$$

where

$$g'_{\mu\nu} = g_{\rho\sigma}(x) \frac{\partial x^\mu}{\partial x'^\rho} \frac{\partial x^\nu}{\partial x'^\sigma} \tag{10.8}$$

describes how to transform the metric tensor from one coordinate system to another. Indeed, this is consistent with the tensor transformation law of Eq. (8.66), which applied in flat space only, such that the only permissible coordinate transformations were Lorentz transformations between inertial frame. Here, this law has been generalised to allow arbitrary coordinate transformations. Furthermore, Eq. (10.8) tells us that the explicit form of the metric for a particular physical solution can be written in infinitely many ways, by performing arbitrary coordinate transformations. This is reminiscent of gauge transformations in electromagnetism, where the gauge field $A_\mu$ takes a different form in different gauges. This analogy can be taken much further, so that we can even think of gravity as a type of gauge theory, but where the "local gauge transformations" are coordinate transformations.

To make the idea of GR precise, we have to give an equation that says, at least in principle, how the metric of spacetime depends upon the matter and/or energy content within it. This is called the *Einstein field equation*, and we refer the reader to textbooks on GR for further details. Here, we simply note that the Einstein equation can be viewed as a nonlinear field equation for $g_{\mu\nu}(x)$, and it is convenient to decompose this into the metric that would be present in flat space, $\eta_{\mu\nu}$, plus a correction:

$$g_{\mu\nu} = \eta_{\mu\nu} + \kappa h_{\mu\nu}. \tag{10.9}$$

We have introduced a conventional factor of $\kappa = \sqrt{16\pi G_N}$, which otherwise could be absorbed into $h_{\mu\nu}$. Note that the decomposition of Eq. (10.9) can always be written, i.e. it simply defines $h_{\mu\nu}$ in terms of the full metric $g_{\mu\nu}$. However, it becomes particularly useful for weak gravitational fields, in which the second term on the right-hand side is a small correction to the first. We can then substitute Eq. (10.9) into the Einstein equation, and keep only those terms which are linear in the field. Note that from Eq. (10.8), the field $h_{\mu\nu}$ will transform under local coordinate transformations as does the full metric. Thus, an explicit form for $h_{\mu\nu}$ for a given physical solution will depend on the choice of coordinates, or *gauge*, and we can fix this by imposing certain conditions on the field $h_{\mu\nu}(x)$. For vacuum solutions (no matter present) at linear order in $h_{\mu\nu}$, a particularly convenient choice is the so-called *transverse traceless* (*TT*) *gauge*, defined by the conditions

$$\eta^{\mu\nu} h_{\mu\nu} = 0, \quad \partial^\mu h_{\mu\nu} = 0. \tag{10.10}$$

Then the Einstein equation to linear order in $h_{\mu\nu}$ reduces to

$$\left[\frac{\partial^2}{\partial t^2} - \nabla^2\right] h_{\mu\nu} = 0, \tag{10.11}$$

which we can recognise as the wave equation.[3] The solutions are *gravitational waves*, which are ripples in the fabric of spacetime that travel at the

[3]Recall that we are in natural units. If not, there is an additional factor of $1/c^2$ in the first term of Eq. (10.11).

speed of light.[4] Gravitational waves can be generated by the collision of heavy objects such as neutron stars or black holes, and were first detected directly by the LIGO experiment in 2016, almost exactly a century after they were first predicted!

In the previous chapter, we saw that wherever a field theory has wave-like solutions, we can consider the quantum version of the theory, in which these waves come in discrete lumps or quanta, which correspond to the particles associated with the field. The same is true in gravity, and quanta of the field $h_{\mu\nu}$ are known as *gravitons*. Likewise, we will refer to $h_{\mu\nu}$ as the *graviton field* from now on. In principle, one may set up the full machinery of quantum field theory for GR, including scattering amplitudes for gravitons, with associated Feynman diagrams and rules. However, this turns out to be a lot more complicated than in Yang–Mills theory. As an example, the standard form for the vertex factor describing three interacting gluons has six mathematical terms in it, whereas the similar result for three gravitons has over 170! It is thus not at all clear that (quantum) results in GR should have anything to do with those in (non-)abelian gauge theories. For scattering amplitudes, however, there is a remarkable relation between Yang–Mills and gravity theories, which we explore in the following section.

## 10.3 The Double Copy for Scattering Amplitudes

As hinted at in the previous section, if one calculates scattering amplitudes in quantum gravity, the results are horrendously complicated. However, this complication is largely associated with intermediate steps of the calculation, due to the huge amount of redundancy in the theory corresponding to the fact that $h_{\mu\nu}$ is only defined up to local coordinate transformations. When final results for amplitudes are assembled, they are much simpler, and indeed start to look much more like their gauge theory counterparts. This idea was taken further in 2012 by Bern, Carrasco and Johansson who formulated an idea known as the *double copy*. It states that it is possible to choose the kinematic numerators $\{n_i\}$ in the gauge theory amplitude of Eq. (10.2) in such a way that the formula

$$\mathcal{M}_m^{(L)} \sim \left(\frac{\kappa}{2}\right)^{m-2+2L} \sum_i \left(\prod_{i=1}^{L} \int d^d k_i\right) \frac{n_i \tilde{n}_i}{\prod_\alpha p_\alpha^2} \qquad (10.12)$$

[4] Note that it is equally valid to say that light moves at the speed of gravity. The constant $c$ acts as a universal limiting speed in nature, that it is impossible to exceed.

is automatically a gravity amplitude with $L$ loops and $m$ external legs. To go from Eq. (10.2) to Eq. (10.12), the gauge theory coupling constant has been replaced with its gravitational counterpart appearing in Eq. (10.9), up to numerical factors. Furthermore, the colour information has been stripped off, and replaced with a second set of kinematic numerators $\{\tilde{n}_i\}$. This might come from the same gauge theory as the one we started with, or they might come from a different gauge theory. Different choices of gauge theory numerators give amplitudes in different gravity theories, consisting of gravity coupled to additional particles. In particular, choosing pure Yang–Mills theory for both sets of gauge theory numerators gives rise to GR plus two additional spinless fields (called the axion and dilaton).

The procedure outlined above is called the *double copy*, due to the fact that two copies of the gauge theory numerators are selected in creating a gravity amplitude. We could also have gone the other way, and replaced the kinematic numerators in Eq. (10.2) with a second set of colour information. This is called the *zeroth copy*, and the resulting amplitude turns out to describe scattering of scalar particles that have two different types of colour charge. The relevant theory has become known as *biadjoint scalar theory* and, although it is not a physical theory of nature itself, its structure and dynamics clearly has a role to play in understanding amplitudes in gauge and gravity theories.

The double copy remains somewhat mysterious, but suggests a profund relationship between three different types of theory, as summarised in Figure 10.3. The obvious questions to ask are: (i) where do the double copy and its related correspondences come from? (ii) are these relationships true only for scattering amplitudes in perturbation theory, or do they indicate a genuinely deep connection between our theories of nature? If the latter is true, it suggests our century-old way of approaching quantum

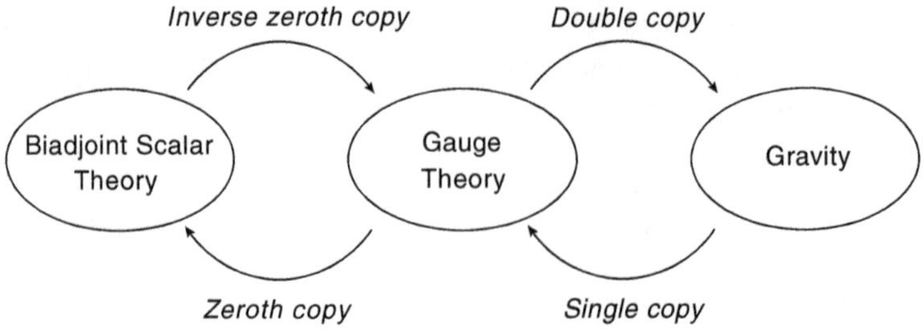


Fig. 10.3 Relationships between various types of theory.

field theory has been hiding a common underlying structure, whose most extreme implication is that the rules of QFT need to be rewritten! Even if only certain quantities turn out to be related between the theories shown in Figure 10.3, it is still highly useful to know what these are. Calculations in even classical gravity are hugely complex, so that new calculational tools are always needed. The double copy provides such a tool in principle: it allows us to "recycle" simpler gauge theory results, thus arriving at new results in gravity a lot more efficiently than using traditional methods. Indeed, one of the earliest uses of the double copy was to find new evidence that a modified version of GR — known as $\mathcal{N} = 8$ *supergravity* — was better behaved than expected, and thus might correspond to a consistent field theory of quantum gravity.

Regarding the origin of the double copy, there is a partial explanation of where it comes from, that works for amplitudes with no loops (i.e. tree-level Feynman diagrams only). The explanation relies on *string theory*, a hypothetical theory of nature in which the point-like particles are replaced by string-like objects, which may be open (i.e. with two endpoints), or closed to form a loop. At low energies, string theories look like field theories, where open strings give rise to non-abelian gauge theories, and closed strings give rise to gravitons and related particles. In the 1970s, Kawai, Lewellen and Tye (KLT) found that scattering amplitudes for closed strings can be written as certain sums of products of amplitudes for open strings. Known as the *KLT relations*, these give rise, at low energy, to precisely the double copy for field theory amplitudes. However, the KLT relations do not obviously generalise to loop-level amplitudes in string theory, whereas the double copy has been tested to multiple loop order in some cases. Hence, although there is overlap between the KLT relations and the double copy, they are not necessarily the same thing.

## 10.4 The Double Copy for Classical Solutions

Much of the research activity since the discovery of the double copy has been to try to ascertain whether or not it generalises from perturbative scattering amplitudes. In other words, should we take Figure 10.3 as applying to the complete theories, whatever this means? If so, we ought to be able to find arbitrary quantities in the different theories, and match them up with each other. The first such quantities to be considered were exact classical solutions of GR and gauge theory. It is still not known how to double copy arbitrary exact solutions, even though this would be highly useful. For example, it could be used to generate *new* solutions of General Relativity, which are very difficult to find in general. However, there is a special family of exact solutions where the double copy is very precisely understood. Known as *Kerr–Schild solutions* in GR, they are such that the graviton decomposes as follows:

$$h_{\mu\nu}(x) = \phi(x) k_\mu k_\nu. \tag{10.13}$$

Here $\phi(x)$ is a scalar field, and $k_\mu$ a vector field. The latter cannot be arbitrary, but must satisfy the following conditions:

$$\eta_{\mu\nu} k^\mu k^\nu = 0, \quad k_\mu \partial^\mu k_\nu = 0, \tag{10.14}$$

where the latter condition holds for all values of the index $\nu$. Upon substituting Eq. (10.13) (via Eq. (10.9)) into the Einstein equations, these become linear, and thus easier to solve. Furthermore, any Kerr–Schild solution is an exact solution, potentially allowing exact statements to be made about the double copy. To this end, one can show that, for any stationary (time-independent) Kerr–Schild solution, the gauge field

$$A_\mu = \phi k_\mu \tag{10.15}$$

satisfies the Maxwell equations. To obtain this, we have stripped off one of the Kerr–Schild vectors $k_\mu$ from Eq. (10.13), to give the right number of spacetime indices for an electromagnetic field. Repeating this procedure gives the field $\phi$ itself, which turns out to satisfy the biadjoint scalar field equation. In general, the single copy of a gravity solution should be a solution to Yang–Mills theory, which is nonlinear (as is biadjoint scalar theory). However, in all three theories considered here, the field equations turn out to linearise, such that the gauge theory is equivalent to electromagnetism. The Kerr–Schild procedure thus gives a concrete realisation of Figure 10.3 for exact classical solutions.

Although the above discussion is rather abstract, the Kerr–Schild double copy applies to some of the most famous gravitational objects in our universe. Arguably the most well-known solution of GR is the *Schwarzschild black hole*, a spherically symmetric solution for the metric around a point mass $M$ (or other spherically symmetric mass distribution). At a distance

$$r_* = 2G_N M, \tag{10.16}$$

from the mass, known as the *Schwarzschild radius*, there is a spherical surface known as the *horizon*. Anything that enters the horizon — including light itself — is unable to escape, hence the term *black hole*. The Kerr–Schild form of the solution is given by Eqs. (10.9) and (10.13), with

$$\phi = \frac{\kappa}{2}\frac{M}{4\pi r}, \quad k^{\mu} = (1, 1, 0, 0), \tag{10.17}$$

where we have used spherical polar coordinates such that $x^{\mu} = (t, r, \theta, \phi)$. Upon taking the single copy of this solution, one finds a gauge field

$$A_{\mu} = \left(\frac{Qe}{4\pi r}, \frac{Qe}{4\pi r}, 0, 0\right), \tag{10.18}$$

again in spherical polars, where we have relabelled

$$\frac{\kappa}{2} \to e, \quad M \to Q, \tag{10.19}$$

where $e$ is the coupling constant of electromagnetism, and the second replacement will become clear shortly. The gauge field of Eq. (10.18) may not look immediately recognisable, but one may show that there is a gauge transformation (Eq. (9.7)) that takes it into the form

$$A_{\mu} = \left(\frac{Qe}{4\pi r}, 0, 0, 0\right). \tag{10.20}$$

Comparing with Eq. (8.62) (and recalling that we are in natural units such that $\epsilon_0$ and $c$ are absent), we can see that there is zero magnetic vector potential. Furthermore, comparison with Eq. (3.29) reveals that the electrostatic potential corresponds to that of a point charge (of charge $Q$ in units of the basic unit of charge $e$), which in turn reproduces Coulomb's law. Thus we find that the single copy of Figure 10.3 takes a black hole in GR, and turns it into a simple point charge in gauge theory!

At first glance, the above story appears to have nothing to do with the double copy for scattering amplitudes that we discussed in Section 10.3. However, there are distinct similarities. Firstly, the amplitudes double copy entailed replacing colour (charge) information in the gauge theory, with kinematic information (e.g. momentum) in the gravity theory. This is precisely what happens for the Schwarzschild black hole: a static charge gets replaced by a mass, which is a kinematic property. Secondly, in the gravity amplitude of Eq. (10.12), the numerator factors and coupling change with respect to Eq. (10.2), but the denominator factors do not change. This is analogous to how, in the Kerr–Schild double copy, the vectors $k_{\mu}$ and charge get relabelled or modified, but the field $\phi$ stays intact. Indeed, the fact that the field $\phi$ is somehow a classical counterpart of the denominators in the amplitude formula can be made much more precise, although we will not do so here.

The above suggestive features of the Schwarzschild black hole turn out to be true for other classical solutions. For example, a more complicated — and more physically relevant — example is the *Kerr black hole*, that corresponds to a rotating disk of mass in gravity. Its single copy turns out to be a rotating disk of charge, whose distribution is closely related to the mass distribution that is needed in the GR solution. In Figure 10.4, we show the magnetic field of the single copy of the Kerr black hole, where the disk lies horizontally in the centre of the figure, from $x = -1$ to $x = 1$. At short distances, the magnetic field has a complicated profile due to the non-trivial distribution of charge on the disk. At large distances, the field looks like that of a bar magnet. To understand why, note that we saw in Chapter 5 that loops of current look like magnetic dipoles, i.e. like bar

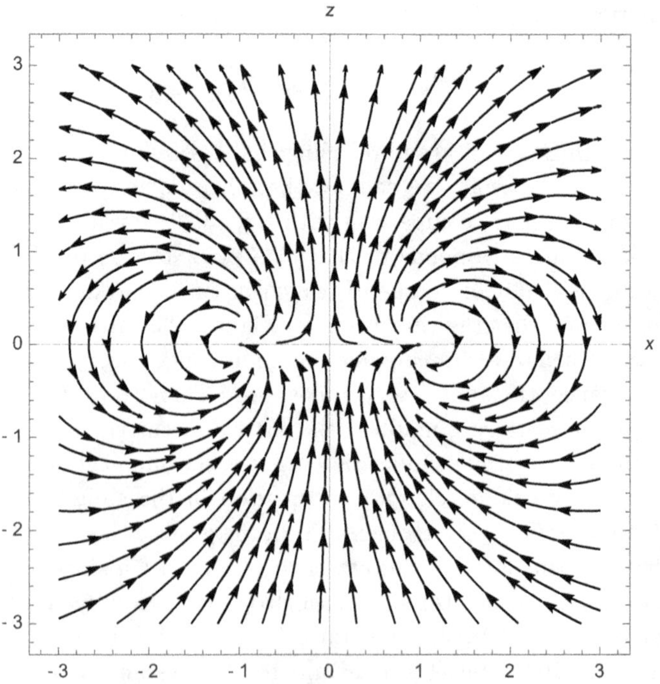


Fig. 10.4 The magnetic field of the single copy of the Kerr black hole, drawn in the $(x, z)$ plane. The rotating disk of charge lies horizontally in the plane $z = 0$, and has unit radius.

magnets. A rotating disk of charge is equivalent to a series of nested current loops, hence gives rise to a dipole field at large distances!

Since the Kerr–Schild procedure for double-copying exact classical solutions, a number of other exact double copies have appeared in the research literature. All of them have in common that how charge behaves in the gauge theory is somehow mapped to how mass (or energy or momentum) behaves in the gravity theory. This indicates that much of the complicated dynamics in gravity can actually be understood in terms of a much simpler theory, at least for the special cases in which exact classical double copies can be applied. For more complicated solutions, we can solve the equations of motion in each theory perturbatively as an expansion in the coupling constant, just as we do for scattering amplitudes. We increasingly understand how to double copy classical solutions of a gauge theory order-by-order in the coupling, and this has in turn generated new results relevant for gravitational wave experiments. The following years are likely to see great progress on two fronts: (i) a conceptual understanding of where the full double copy comes from, and also what it is trying to tell us about the foundations of quantum field theory; (ii) practical applications of the double copy to astrophysics or cosmology, allowing new and efficient ways to solve complicated gravitational problems.

This is perhaps a fitting place to end this book. We started by assuming no prior knowledge of electromagnetism, and have finished by showing the reader that this subject is much closer to contemporary research in both theoretical and applied physics than perhaps they thought. The importance of electromagnetism as a cornerstone of modern physics is indisputable, and the author humbly hopes that the reader will take up some of the challenges raised in this book in their own scientific career!

## Exercises

(1) A standard result in GR states that, for sufficiently weak gravitational fields, the Newtonian gravitational potential $V_G$ can be obtained from the metric tensor. In natural units, this relation can be written as

$$g_{00} \simeq 1 + 2V_G.$$

Show that, for the Kerr–Schild solution of Eq. (10.17), one obtains the Newtonian potential

$$V_G = \frac{G_N M}{r},$$

which is the correctly normalised potential for a point charge in Newton's theory of gravity.

(2) How do the replacements in Eq. (10.19) compare with the double copy for scattering amplitudes, of Eqs. (10.2) and (10.12)?

(3) Show that the gauge transformation of Eq. (9.7), with

$$\alpha = \frac{Qe^2}{4\pi}\log\left(\frac{r}{r_0}\right)$$

(for some constant $r_0$) can be used to convert the single copy of the Schwarzschild black hole, Eq. (10.18), into the more recognisable form of Eq. (10.20). You may use the fact that the four-dimensional gradient operator in spherical polar coordinates $(t, r, \theta, \phi)$ is given by

$$\partial_\mu = \left(\frac{\partial}{\partial t}, \frac{\partial}{\partial r}, \frac{1}{r}\frac{\partial}{\partial \theta}, \frac{1}{r\sin\theta}\frac{\partial}{\partial \phi}\right).$$

(4) In Figure 10.4, we show the magnetic field of the single copy of the Kerr black hole, which at large distances looks like a magnetic dipole. What will the electric field look like at large distances?

# Appendix A

# Line and Surface Integrals

Throughout this book, there are various instances of line and surface integrals. However, in all of the cases considered, there is some symmetry property or specialness, such that the full general machinery of how to carry out line or surface integrals is never needed. The aim of this appendix is to explain the general case in detail, to complete the picture. Let us first discuss line integrals.

## A.1 Line Integrals

Consider a vector field $\boldsymbol{F}(\boldsymbol{x})$, namely a function that associates a vector with every single point in space. Note that, although vector fields may also depend on time in general, this will not be important in what follows, so that we will only denote the dependence of the vector on spatial position $\boldsymbol{x}$. Following the example of Eq. (3.20), we may write a general form for the line integral of a vector field:

$$I = \int_C \boldsymbol{F} \cdot d\boldsymbol{x}. \tag{A.1}$$

As often happens in mathematics, this notation does not by itself define what we mean, without much further explanation! In particular, it is understood that we have a curve $C$ in mind, that begins and ends somewhere in space, albeit with the endpoints possibly being infinitely far away. The integral over positions $\boldsymbol{x}$ is then taken to be only over those positions that lie on the curve $C$. An example is shown in Figure A.1, which shows a position vector extending from the origin to a given point on some curve $C$. Note that the curve is a *1D* object. In other words, we need a single parameter $\lambda$ to measure where we are on $C$. This might, for example, be

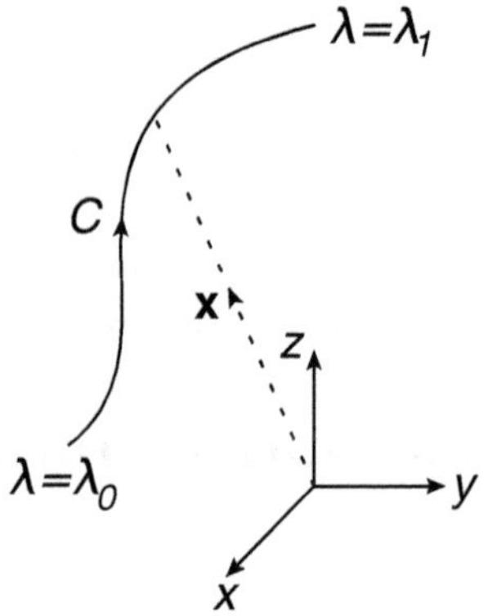


Fig. A.1 A curve $C$ in space can be specified using a parameter $\lambda_0 \leq \lambda \leq \lambda_1$, such that $\boldsymbol{x}(\lambda)$ is a position vector from the origin to a given point on the curve.

simply the length along the curve in some appropriate units. But it might also be something more complicated: we can choose any function of $\lambda$ that is always increasing along the curve, and is single-valued, so that a given value of $\lambda$ always labels a unique point on $C$. Let us thus be fully general, and say that $\lambda = \lambda_0$ at the start of the curve, and $\lambda = \lambda_1$ at the end of the curve, as shown in Figure A.1. A given point on the curve will then be given by

$$\boldsymbol{x}(\lambda), \quad \lambda_0 \leq \lambda \leq \lambda_1. \tag{A.2}$$

This is called a *parametrisation* of the curve, and it is instructive to illustrate this with a simple example in two dimensions. Consider the line shown in Figure A.2, which goes from the origin to the point (1,1). Considered as a graph, the equation of this line is

$$y = x, \quad 0 \leq x,\ y \leq 1, \tag{A.3}$$

so that a possible parametrisation is

$$\begin{pmatrix} x \\ y \end{pmatrix} = \begin{pmatrix} \lambda \\ \lambda \end{pmatrix}, \quad 0 \leq \lambda \leq 1. \tag{A.4}$$

This parametrisation does two things: it incorporates the fact that $x$ and $y$ must be equal, and it also reproduces the range of $x$ and $y$, given the range of $\lambda$. Furthermore, Eq. (A.4) has exactly the form of Eq. (A.2), in that it gives the position vector of points along the line in terms of $\lambda$. We see directly from the right-hand side of Eq. (A.4) that $x$ and $y$ are not

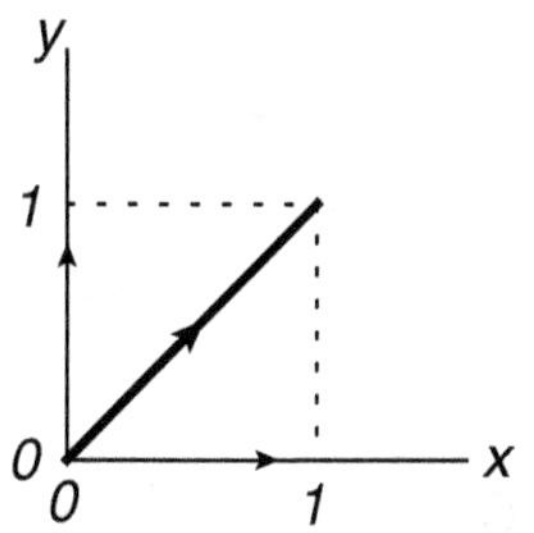


Fig. A.2 A line in the $(x, y)$ plane.

independent, in that they both depend on the *same* parameter $\lambda$. This is also made clear by Eq. (A.3), which tells us explicitly how $x$ and $y$ are related.

Returning to the case of a general curve $C$, a parametrisation such as Eq. (A.4) allows us to turn the line integral of Eq. (A.1) into a scalar integral, which we can then carry out. First, let us understand the meaning of $d\boldsymbol{x}$: if $\boldsymbol{x}$ is the position vector of a point on $C$, then $d\boldsymbol{x}$ represents the difference between the position vectors of two points which are infinitesimally close on the curve. By the rules of vector subtraction given in Chapter 2, this will be a vector that connects the two points on the curve. As these are taken infinitesimally close to each other, $d\boldsymbol{x}$ points along the curve itself, becoming tangent to it (Figure A.3). Having interpreted $d\boldsymbol{x}$, let us now use the chain rule of differentiation to write

$$d\boldsymbol{x} = \frac{d\boldsymbol{x}}{d\lambda} d\lambda. \tag{A.5}$$

If you have not seen the chain rule applied to a vector before, don't worry: we can view this as three separate equations, one for each component, such that we have (in, e.g. three dimensions)

$$dx = \frac{dx}{d\lambda} d\lambda, \quad dy = \frac{dy}{d\lambda} d\lambda, \quad dz = \frac{dz}{d\lambda} d\lambda. \tag{A.6}$$

Here $(dx, dy, dz)$ are the changes in the coordinates as we move along the curve, such that these changes are not independent from each other. Equation (A.5) contains the vector

$$\frac{d\boldsymbol{x}}{d\lambda}, \tag{A.7}$$

which is no longer infinitesimally small, and points along the curve. To see the latter, note that multiplying or dividing a vector by a scalar does

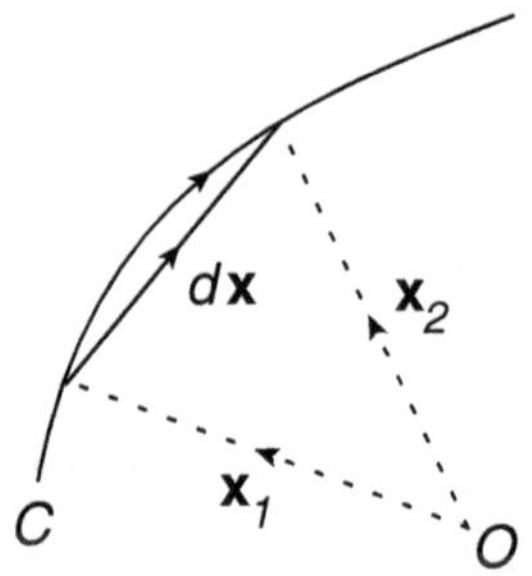


Fig. A.3 Consider two points on a curve $C$, at positions $\boldsymbol{x}_1$ and $\boldsymbol{x}_2$. The difference $d\boldsymbol{x} = \boldsymbol{x}_2 - \boldsymbol{x}_1$ points from the first point to the second, such that when the points become infinitesimally close, $d\boldsymbol{x}$ is tangent to the curve.

not change its direction. The vector $d\boldsymbol{x}$ points along the curve, so that "dividing" by the small parameter length $d\lambda$ (which one does in defining the derivative as a formal limit when $d\lambda \to 0$) does not change the direction. It is common to call Eq. (A.7) the *tangent vector to the curve*, although its length will clearly depend on the parametrisation used. For a given choice of parametrisation, the tangent vector at each point along the curve will be unique.

Substituting Eq. (A.5) into Eq. (A.1), the latter becomes

$$I = \int_{\lambda_0}^{\lambda_1} \left( \boldsymbol{F}(\boldsymbol{x}(\lambda)) \cdot \frac{d\boldsymbol{x}}{d\lambda} \right) d\lambda. \tag{A.8}$$

This is a normal scalar integral, where we integrate over the various values the parameter $\lambda$ can take, and where the limits of integration correspond to the endpoints of the curve $C$. The integrand contains the dot product of the vector field at each point along the curve, with the tangent vector. Given that we know $\boldsymbol{x}$ as a function of $\lambda$, this integrand can be expressed purely in terms of the single parameter $\lambda$. Note that, whilst the integrand in Eq. (A.8) depends on the particular parametrisation that is used, the total line integral does not. This is hopefully clear from Eq. (A.1), which shows that the formal dependence on $\lambda$ cancels when one writes the integral purely in terms of the position vector $\boldsymbol{x}$.

In summary, the way to carry out a line integral is as follows: (i) find a parametrisation of the curve $C$, involving some parameter $\lambda$; (ii) for your given parametrisation, rewrite the integral to involve the tangent vector $d\boldsymbol{x}/d\lambda$, and express the integrand of the line integral purely in terms of $\lambda$; (iii) carry out the integral in $\lambda$ as a conventional scalar integral, where the

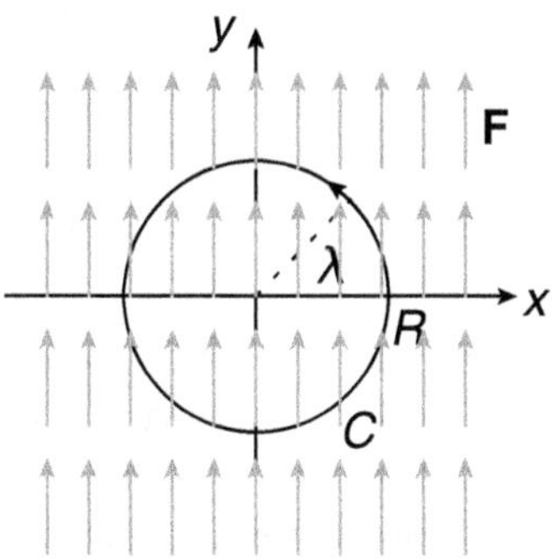


Fig. A.4 A circular curve of radius $R$ in the $(x, y)$ plane, and a constant vector field in the $y$-direction.

limits of integration are set by the parameter values corresponding to the endpoints of the curve. As an example, consider a constant 2D vector field

$$\boldsymbol{F} = \begin{pmatrix} 0 \\ \alpha \end{pmatrix} \tag{A.9}$$

in the $y$-direction, and the closed curve shown in Figure A.4, consisting of a circle of radius $R$ in the $(x, y)$ plane. To parametrise the curve, we need a single parameter that takes us along the curve, and for which we can write explicit equations for $x$ and $y$. It is natural to choose $\lambda$ to correspond to the angle measured anti-clockwise from the positive $x$-axis, as shown in the figure. We can then write

$$\boldsymbol{x} = \begin{pmatrix} x \\ y \end{pmatrix} = \begin{pmatrix} R\cos\lambda \\ R\sin\lambda \end{pmatrix}, \quad 0 \leq \lambda \leq 2\pi. \tag{A.10}$$

Although it looks as if there are two parameters on the right-hand side, the radius $R$ is fixed, so that only $\lambda$ is varying. The tangent vector to the curve is then given by

$$\frac{d\boldsymbol{x}}{d\lambda} = \frac{d}{d\lambda}\begin{pmatrix} R\cos\lambda \\ R\sin\lambda \end{pmatrix} = \begin{pmatrix} -R\sin\lambda \\ R\cos\lambda \end{pmatrix}, \tag{A.11}$$

such that the integrand of the line integral becomes

$$\boldsymbol{F} \cdot \frac{d\boldsymbol{x}}{d\lambda} = \begin{pmatrix} 0 \\ \alpha \end{pmatrix} \cdot \begin{pmatrix} -R\sin\lambda \\ R\cos\lambda \end{pmatrix} = \alpha R\cos\lambda. \tag{A.12}$$

The line integral of the vector field along the whole curve $C$ is then given by

$$\int_C \boldsymbol{F}\cdot d\boldsymbol{x} = \int_0^{2\pi} \alpha R\cos\lambda\, d\lambda = \Big[\alpha R\sin\lambda\Big]_0^{2\pi} = 0. \qquad \text{(A.13)}$$

This turns out to be zero, which in fact can be straightforwardly understood from Figure A.4: the vector field has a non-zero component *along* the curve on the right-hand side ($x > 0$), but *against* the curve on the left-hand side ($x < 0$). By the symmetry of the figure, the corresponding contributions to the line integral must exactly cancel out. We learn a valuable lesson here, namely that it is always worth looking for symmetry arguments, or any other considerations, that can be used to avoid doing unnecessary integrals!

## A.2 Surface Integrals

Throughout the book, we have also encountered the notion of the surface integral, or flux, of a vector field $\boldsymbol{F}(\boldsymbol{x})$, and the aim of this section is to explain how surface integrals can be constructed in full generality, for arbitrary surfaces in 3D. First note that, in contrast to the one-dimensional lines considered in the previous section, surfaces are two-dimensional: we need two parameters to specify a point on the surface. In principle, one may completely determine a given surface $S$ in three dimensions by an equation of the form

$$u(x, y, z) = 0. \qquad \text{(A.14)}$$

To see this, note that there are three independent components in a 3D space, namely $x$, $y$ and $z$. Equation (A.14) implies a single constraint linking these coordinates, so that only two independent degrees of freedom remain, as required. We have used Cartesian coordinates in Eq. (A.14), but this is not necessary. We could also write Eq. (A.14) as $u(\boldsymbol{x}) = 0$, where $\boldsymbol{x}$ is the position vector in an arbitrary coordinate system, and similar reasoning would then apply.

Given a vector field $\boldsymbol{F}(\boldsymbol{x})$ and a surface $S$, the general definition of a surface integral is

$$J = \int\!\!\int_S \boldsymbol{F}\cdot d\boldsymbol{S}. \qquad \text{(A.15)}$$

As explained in Chapter 3, this is to be interpreted as follows: we can divide $S$ into infinitesimal segments. On each segment, we can consider the vector area $d\boldsymbol{S}$ pointing out from the surface, whose magnitude is the area of the segment. We then take the dot product of this vector area with the vector field, which will be constant on a given segment given that this is infinitesimally small. Finally, we sum over all the possible segments, which is a continuous sum (an integral) given the continuous nature of the parameters labelling where we are on the surface. Let us first see how to construct the vector area, and we may write this for a particular segment of the surface as

$$d\boldsymbol{S} = \hat{\boldsymbol{n}} dS, \tag{A.16}$$

where $\hat{\boldsymbol{n}}$ is a unit vector normal to the surface, and $dS$ the (scalar) area of the segment. It may seem impossible that we can find a unit normnal vector of an arbitrary surface $S$ at every point on the surface, but it turns out that there is a remarkably simple procedure for doing this, given the defining property for our surface $S$ of Eq. (A.14). If we consider the gradient operator introduced in Section 3.34, we may operate it on the function $u(x, y, z)$ defining $S$, to get

$$\nabla u = \begin{pmatrix} \dfrac{\partial u}{\partial x} \\ \dfrac{\partial u}{\partial y} \\ \dfrac{\partial u}{\partial z} \end{pmatrix}. \tag{A.17}$$

At our given segment of the surface, we may consider a small displacement $d\boldsymbol{x} = (dx, dy, dz)$, such that

$$(\nabla u) \cdot d\boldsymbol{x} = \frac{\partial u}{\partial x} dx + \frac{\partial u}{\partial y} dy + \frac{\partial u}{\partial z} dz = du, \tag{A.18}$$

where we have used the chain rule of partial differentiation in the second equality. We thus find that the vector gradient of a function, dotted with a displacement $d\boldsymbol{x}$, tells us about the change in the function in the direction of $d\boldsymbol{x}$, and indeed this is true for *any* function, not just functions that define surfaces. Returning to the case of our surface segment, however, we know from Eq. (A.14) that $u(x, y, z)$ does not change if we move parallel to the surface. That is, we must have

$$(\nabla u) \cdot d\boldsymbol{x} = 0$$

for any displacements lying in the surface $S$. This in turn implies that $\nabla u$ must be perpendicular to the surface, and thus a unit normal vector to *any* surface defined by a constraint equation such as Eq. (A.14) is given by

$$\hat{\boldsymbol{n}} = \frac{\nabla u}{|\nabla u|}. \tag{A.19}$$

All quantities in this equation are understood to be evaluated at a particular point on the surface $S$. Given that the function $f(x, y, z)$ depends upon position, it is clear that the unit normal vector to the surface will be different in general at different points on the surface, which is indeed true for curved surfaces![1]

Having found a normal vector to the surface, let us now see how to find the area of a given segment, $dS$. In some cases, we can use special coordinate systems to do this directly. However, there is a general method that will always work in principle, but which may prove complicated in practice. Let us consider a segment of a general surface $S$ lying above the $(x, y)$ plane, as shown in Figure A.5. This has a projected area in the $(x, y)$ plane of

$$dS' = dx\, dy,$$

where the appropriate distances on the right-hand side are labelled in the figure. Using trigonometry, however, this must be related to the area of the segment $dS$ by

$$dS' = dS \cos\theta,$$

where $\theta$ is the angle shown. It is the angle between the unit normal to the segment, and the unit vector $\hat{\boldsymbol{k}}$ in the $z$-direction, such that one has

$$\cos\theta = \hat{\boldsymbol{n}} \cdot \hat{\boldsymbol{k}},$$

[1]Note that one may consider one of two different directions for $\hat{\boldsymbol{n}}$, given that there are two directions that are orthogonal to the surface. Mathematically, this means that we can take Eq. (A.19) to be true only up to an overall minus sign on the right-hand side. To fix conventions for a closed surface, we can choose that the normal always points outwards. For an open surface, one way to fix the direction of the normal is to introduce an orientation on the curve that bounds the surface, and then use a right-hand rule.

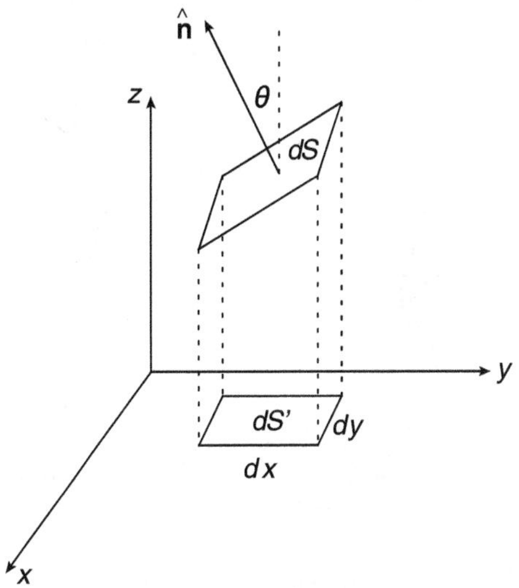


Fig. A.5 A segment of a surface $S$, that lies above the $(x, y)$ plane (rest of the surface not shown).

where we have used Eq. (2.9), plus the fact that both $\hat{\boldsymbol{n}}$ and $\hat{\boldsymbol{k}}$ are unit vectors such that $|\hat{\boldsymbol{n}}| = |\hat{\boldsymbol{k}}| = 1$. Putting things together, we find

$$dS = \frac{|\nabla u| dx\, dy}{\nabla u \cdot \hat{\boldsymbol{k}}} = \frac{|\nabla u| dx\, dy}{\partial u / \partial z}. \tag{A.20}$$

The vector area on a given segment is thus given by

$$d\boldsymbol{S} = \hat{\boldsymbol{n}} dS = \frac{\nabla u}{|\nabla u|} \frac{|\nabla u| dx\, dy}{\partial u / \partial z} = \frac{(\nabla u) dx\, dy}{\partial u / \partial z}. \tag{A.21}$$

For a given vector field $\boldsymbol{F}(\boldsymbol{x})$, we can now evaluate the surface integral of Eq. (A.15): substituting Eq. (A.21) yields

$$J = \int dx \int dy \, \frac{\boldsymbol{F} \cdot \nabla u}{\partial u / \partial z}, \tag{A.22}$$

which becomes a conventional 2D integral in the $(x, y)$ plane. We have not written explicit limits on the $x$ and $y$ integrals, but these will be given by the projection of the boundary of the surface $S$ onto the $(x, y)$ plane. Furthermore, it is understood that all quantities in the integrand are evaluated on the surface $S$ itself: the projection into the $(x, y)$ plane is purely so that we can find an expression for the area of each segment $dS$. The coordinates

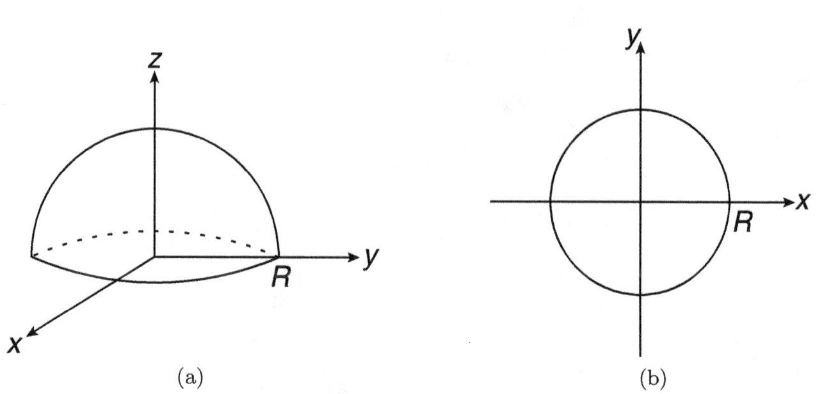


Fig. A.6 (a) A hemispherical surface whose base is a circle of radius $R$ in the $(x, y)$ plane; (b) projection onto the $(x, y)$ plane.

$x$ and $y$ then label points on the surface, where the appropriate point for a given $(x, y)$ can be found by going vertically upwards.

As an example, consider the flux of the vector field

$$\boldsymbol{F}(\boldsymbol{x}) = \alpha\left(\frac{x}{r}, \frac{y}{r}, \frac{z}{r}\right), \quad r = \sqrt{x^2 + y^2 + z^2}, \tag{A.23}$$

through the hemispherical surface shown in Figure A.6(a). Note that we will take the example of an open surface, such that the base of this hemisphere is not included in the surface. Then, we can write a constraint equation for the surface as in Eq. (A.14):

$$u(x, y, z) = x^2 + y^2 + z^2 - R^2 = 0, \quad z \geq 0, \tag{A.24}$$

from which we find

$$\nabla u = (2x, 2y, 2z)\,, \tag{A.25}$$

where the third component constitutes $\partial u/\partial z$. We can then evaluate the integrand of Eq. (A.22), which gives

$$\begin{aligned}\frac{\boldsymbol{F}\cdot\nabla u}{\partial u/\partial z} &= \frac{\alpha}{2z}\left(\frac{x}{r}, \frac{y}{r}, \frac{z}{r}\right)\cdot(2x, 2y, 2z)\\ &= \frac{\alpha(x^2 + y^2 + z^2)}{zr}.\end{aligned}$$

This is to be evaluated on the surface, for which Eq. (A.24) gives

$$r = R, \quad z = \sqrt{R^2 - x^2 - y^2},$$

so that we have

$$\frac{\boldsymbol{F}\cdot\nabla u}{\partial u/\partial z} = \frac{\alpha R}{\sqrt{R^2 - x^2 - y^2}}.$$

Equation (A.22) then becomes

$$\begin{aligned}
&\int_{-R}^{R} dx \int_{-\sqrt{R^2-x^2}}^{\sqrt{R^2-x^2}} dy \, \frac{\alpha R}{\sqrt{R^2 - x^2 - y^2}} \\
&= \alpha R \int_{-R}^{R} dx \left[\tan^{-1}\left(\frac{y}{\sqrt{R^2 - x^2 - y^2}}\right)\right]_{-\sqrt{R^2-x^2}}^{\sqrt{R^2-x^2}} \\
&= \alpha \pi R \int_{-R}^{R} dx \\
&= 2\alpha\pi R^2.
\end{aligned}$$

For this particular example, there is in fact an easier way to obtain this result. First, we can note that our vector field of Eq. (A.23) has the form

$$\boldsymbol{F} = \alpha\hat{\boldsymbol{r}}, \tag{A.26}$$

where $\hat{\boldsymbol{r}}$ is a unit vector in the radial direction. Furthermore, $\hat{\boldsymbol{r}}$ is clearly also the unit normal to the hemisphere, at all points on the surface (n.b. the radial direction changes depending on where we are in space). Thus, we can write

$$\int\int_S \boldsymbol{F}\cdot d\boldsymbol{S} = \int\int_S \boldsymbol{F}\cdot\hat{\boldsymbol{n}}dS = \alpha\int\int_S dS.$$

The right-hand side is simply $\alpha$ multiplying the total area of the surface, which is $2\pi R^2$, in agreement with the above result.

# Appendix B

# The Dirac Equation

In Chapter 9, we have shown how requiring local gauge invariance of the theory of the electron, as expressed by the Dirac equation, leads to the existence of electromagnetism. In this appendix, we provide details of where the Dirac equation comes from, and will rely on more knowledge of quantum mechanics than is assumed throughout the rest of this book.

The Dirac equation has its origin in attempts to formulate a relativistic generalisation of the non-relativistic quantum mechanics that was pioneered in the early part of the twentieth century, and which occupies the lower-left panel in Table 9.1. The latter posits the existence of a *wavefunction* $\psi$, satisfying the *Schrödinger equation*

$$\left[-\frac{\hbar^2}{2m}\nabla^2 + V(\boldsymbol{x})\right]\psi(\boldsymbol{x},t) = i\hbar\frac{\partial\psi}{\partial t}, \tag{B.1}$$

where we have taken the example of a single particle with potential energy $V(\boldsymbol{x})$. The wavefunction is a complex quantity, such that

$$|\psi(\boldsymbol{x},t)|^2 d^3\boldsymbol{x}, \tag{B.2}$$

represents the probability of finding the particle in the small volume $d^3\boldsymbol{x}$ around position $\boldsymbol{x}$ at time $t$. To motivate where the Schrödinger equation comes from, note that the total energy of a particle in non-relativistic mechanics is given by

$$E = \frac{\boldsymbol{p}^2}{2m} + V(\boldsymbol{x}), \tag{B.3}$$

where the first term is the kinetic energy. Acting on the wavefunction, momentum and energy are represented by the following operators in quantum mechanics:

$$\hat{p} = -i\hbar\nabla, \quad \hat{E} = i\hbar\frac{\partial}{\partial t}. \tag{B.4}$$

Acting on a position-space wavefunction, the potential energy simply becomes the operator

$$\hat{V}(x) = V(x), \tag{B.5}$$

such that replacing each term in Eq. (B.3) with the appropriate operator acting on a wavefunction, one indeed obtains Eq. (B.1).

Early attempts to seek a relativistic generalisation of the Schrödinger equation repeated the above exercise, but using the relativistic energy–momentum relation of Eq. (8.9) for a free particle (with no potential energy). Assuming that there is some appropriate wavefunction $\phi(x)$, where $x$ is the 4-vector position, we obtain

$$\left[\frac{1}{c^2}\frac{\partial^2}{\partial t^2} - \nabla^2 + \frac{m^2c^2}{\hbar^2}\right]\phi(x) = 0. \tag{B.6}$$

This is known as the *Klein–Gordon equation*, although was first considered by Schrödinger, *before* the famous equation above that bears his name! However, attempts to intepret Eq. (B.6) as a relativistic single-particle wave equation ultimately fail. It turns out that if one tries to construct a probability density for the particle, analogous to the non-relativistic case of Eq. (B.2), one finds that this cannot be guaranteed to be positive, as is required. Dirac traced this problem to the fact that Eq. (B.6) is *second-order* in both space and time derivatives, and he thus thought that a sensible relativistic single-particle wave equation could be achieved by requiring that it be *first-order* in all derivatives, such that the energy–momentum relation of Eq. (8.9) is somehow still obeyed. To this end, Dirac considered the following equation:

$$\frac{\hat{E}}{c}\Psi(x) = (\alpha_x\hat{p}_x + \alpha_y\hat{p}_y + \alpha_z\hat{p}_z + \beta mc)\Psi(x), \tag{B.7}$$

where $\hat{E}$ and $\hat{p}_i$ are the usual energy and momentum operators, and $(\alpha_x, \alpha_y, \alpha_z, \beta)$ are to be determined. To do this, one may first rearrange

Eq. (B.7):

$$\left(\frac{\hat{E}}{c} - \alpha_x \hat{p}_x - \alpha_y \hat{p}_y - \alpha_z \hat{p}_z - \beta mc\right) \Psi(x) = 0,$$

before acting with the operator

$$\frac{\hat{E}}{c} + \alpha_x \hat{p}_x + \alpha_y \hat{p}_y + \alpha_z \hat{p}_z + \beta mc$$

to get

$$\left(\frac{\hat{E}^2}{c^2} + \sum_i \alpha_i^2 (\hat{p}_i)^2 + \sum_{i<j} (\alpha_i \alpha_j + \alpha_j \alpha_i) \hat{p}_i \hat{p}_j \right.$$
$$\left. + \sum_i (\alpha_i \beta + \beta \alpha_i) m \hat{p}_i - \beta^2 m^2 c^2 \right) \Psi(x) = 0, \tag{B.8}$$

where we have switched to the index notation $(V_1, V_2, V_3) = (V_x, V_y, V_z)$ for the components of an arbitrary 3-vector $\boldsymbol{V}$. In order to satisfy the quantum version of Eq. (8.9), Eq. (B.8) must be equivalent to

$$\left(\frac{\hat{E}^2}{c^2} - \sum_i \hat{p}_i^2 - m^2 c^2\right) \Psi(x) = 0, \tag{B.9}$$

which thus implies

$$\alpha_i^2 = \beta^2 = 1, \quad \alpha_i \beta + \beta \alpha_i = 0, \tag{B.10}$$

and

$$\alpha_i \alpha_j + \alpha_j \alpha_i = 0, \quad i \neq j. \tag{B.11}$$

Clearly, these conditions cannot be satisfied if $\{\alpha_i\}$ and $\beta$ are numbers. They can instead be matrices, given that multiplication of matrices is not commutative in general. It so happens that the smallest matrices one can consider are $4 \times 4$, in which case the first conditions in Eq. (B.10) are to be interpreted such that the various matrices square to the $4 \times 4$ identity matrix. Upon substituting the form of the operators, Eq. (B.8) becomes

$$\left(\frac{i\hbar}{c}\frac{\partial}{\partial t} + i\hbar \boldsymbol{\alpha} \cdot \nabla - \beta mc\right) \Psi(x) = 0 \tag{B.12}$$

and, for consistency, $\Psi(x)$ must now be regarded as a four-component object, given that it is acted upon by $4 \times 4$ matrices. It is called a (*Dirac*) *spinor*, and we will discuss why it has four components shortly. First, we note that it is convenient to multiply Eq. (B.12) through by the matrix $\beta$ to obtain

$$\left(\frac{i\beta\hbar}{c}\frac{\partial}{\partial t} + i\hbar\beta\boldsymbol{\alpha}\cdot\nabla - \beta mc\right)\Psi(x) = 0.$$

Then one may define the quantity

$$\gamma^\mu = (\beta, \beta\boldsymbol{\alpha}), \tag{B.13}$$

such that Eq. (B.12) may be written (using the Einstein summation convention) as

$$(i\hbar\gamma^\mu\partial_\mu - mc)\Psi(x) = 0. \tag{B.14}$$

As stated in Chapter 9, it is conventional to use so-called *natural units* in particle physics, which amounts to ignoring factors of $c$ and $\hbar$, such that these can be reinstated later using dimensional analysis. Then Eq. (B.14) is precisely the Dirac equation of Eq. (9.2).

The quantities $\gamma^\mu$ are known as *Dirac matrices*, and the conditions of Eqs. (B.10) and (B.11) imply that they obey the relations

$$\gamma^\mu\gamma^\nu + \gamma^\nu\gamma^\mu = 2\eta^{\mu\nu}, \tag{B.15}$$

where there is an implicit $4\times 4$ identity matrix on the right-hand side, and $\eta^{\mu\nu}$ is given in Eq. (8.42). A suitable choice for the Dirac matrices is

$$\gamma^0 = \begin{pmatrix} 1 & 0 & 0 & 0 \\ 0 & 1 & 0 & 0 \\ 0 & 0 & -1 & 0 \\ 0 & 0 & 0 & -1 \end{pmatrix}, \quad \gamma^1 = \begin{pmatrix} 0 & 0 & 0 & 1 \\ 0 & 0 & 1 & 0 \\ 0 & -1 & 0 & 0 \\ -1 & 0 & 0 & 0 \end{pmatrix},$$

$$\gamma^2 = \begin{pmatrix} 0 & 0 & 0 & -i \\ 0 & 0 & i & 0 \\ 0 & i & 0 & 0 \\ -i & 0 & 0 & 0 \end{pmatrix}, \quad \gamma^3 = \begin{pmatrix} 0 & 0 & 1 & 0 \\ 0 & 0 & 0 & -1 \\ -1 & 0 & 0 & 0 \\ 0 & 1 & 0 & 0 \end{pmatrix}. \tag{B.16}$$

Let us now interpret the quantity $\Psi$, which is found to need four components. Given that this theory was found to describe electrons, it makes sense that there are at least two degrees of freedom, given that the electron has spin-1/2. Using standard results from quantum mechanics, this implies that the electron should have two independent spin states, which we can choose to correspond to the electron spin being aligned along the $\pm$ $z$-directions, respectively. However, that does not explain the other two degrees of freedom in $\Psi(x)$, and they turn out to correspond to the existence of *anti-matter*, namely to the fact that the electron has a partner (the positron) with identical properties apart from the sign of the charge. To see why this happens, note that in the non-relativistic energy–momentum relation of Eq. (B.3), the energy is manifestly positive. This is in contrast to the relativistic case of Eq. (8.9), for which one finds two solutions for the energies of particles:

$$E = \pm\sqrt{\boldsymbol{p}^2c^2 + m^2c^4}. \tag{B.17}$$

In a classical theory, one can simply throw away the negative energy solutions: once a particle has positive energy, it will stay that way. In a quantum theory, however, a particle will be able to transition in principle to the states of negative energy, such that these must be taken seriously. Unfortunately, there are potentially infinitely many states of negative energy!

Faced with the problem of negative energy states, it is quantum field theory that comes to the rescue. In that framework, the Dirac equation of Eq. (9.2), rather than being interpreted as a single-particle wave equation, is reinterpreted to have $\Psi(x)$ as a field filling all space. The negative energy particle states get successfully reinterpreted as positive energy *antiparticles*, whose charge is precisely opposite to that of the corresponding particle. Similarly, all the other matter particles in nature have corresponding anti-particles, such that matter and anti-matter can be created in equal amounts. It is for this reason that single particle interpretations of the Klein–Gordon and Dirac equations were doomed to fail: as field theories, they describe many particles, and also creation of both matter and anti-matter particles, limited only by the energy available. It should now hopefully be clear why the Dirac field $\Psi(x)$ has four components: it must describe both spin states of the electron, and both spin states of the anti-electron (the positron).

# Appendix C

# Solutions to Exercises

In this appendix, we present guided solutions for the various exercises that appear at the end of each chapter.

## Chapter 2: Vector Algebra

(1) Using Eq. (2.9), we find that the dot product vanishes if the angle $\theta$ satisfies

$$\cos\theta = 0.$$

Thus, the dot product is zero for vectors that are orthogonal (perpendicular).

For the cross product, Eq. (2.16) tells us that the magnitude is zero if

$$\sin\theta = 0.$$

Thus, the cross-product vanishes if the two vectors are (anti-)parallel.

(2) The basis vectors have Cartesian components

$$\hat{\boldsymbol{\imath}} = \begin{pmatrix} 1 \\ 0 \\ 0 \end{pmatrix}, \quad \hat{\boldsymbol{\jmath}} = \begin{pmatrix} 0 \\ 1 \\ 0 \end{pmatrix}, \quad \hat{\boldsymbol{k}} = \begin{pmatrix} 0 \\ 0 \\ 1 \end{pmatrix}.$$

Then one has

$$\hat{\boldsymbol{\imath}} \times \hat{\boldsymbol{\jmath}} = \begin{pmatrix} 1 \\ 0 \\ 0 \end{pmatrix} \times \begin{pmatrix} 0 \\ 1 \\ 0 \end{pmatrix} = \begin{pmatrix} 0 \times 0 - 0 \times 1 \\ 0 \times 0 - 1 \times 0 \\ 1 \times 1 - 0 \times 0 \end{pmatrix} = \begin{pmatrix} 0 \\ 0 \\ 1 \end{pmatrix} = \hat{\boldsymbol{k}},$$

$$\hat{\boldsymbol{k}} \times \hat{\boldsymbol{\imath}} = \begin{pmatrix} 0 \\ 0 \\ 1 \end{pmatrix} \times \begin{pmatrix} 1 \\ 0 \\ 0 \end{pmatrix} = \begin{pmatrix} 0 \times 0 - 1 \times 0 \\ 1 \times 1 - 0 \times 0 \\ 0 \times 0 - 1 \times 0 \end{pmatrix} = \begin{pmatrix} 0 \\ 1 \\ 0 \end{pmatrix} = \hat{\boldsymbol{\jmath}},$$

$$\hat{\boldsymbol{\jmath}} \times \hat{\boldsymbol{k}} = \begin{pmatrix} 0 \\ 1 \\ 0 \end{pmatrix} \times \begin{pmatrix} 0 \\ 0 \\ 1 \end{pmatrix} = \begin{pmatrix} 1 \times 1 - 0 \times 0 \\ 0 \times 0 - 0 \times 1 \\ 0 \times 0 - 1 \times 0 \end{pmatrix} = \begin{pmatrix} 1 \\ 0 \\ 0 \end{pmatrix} = \hat{\boldsymbol{\imath}},$$

where we have used Eq. (2.21). The remaining results follow from the fact that the cross product is antisymmetric under interchange of the first and second vector.

(3) First, we need the magnitudes of each vector. These are given by

$$|\boldsymbol{a}| = |\boldsymbol{b}| = \sqrt{1^2 + 2^2 + 3^2} = \sqrt{14}.$$

From the definition of the dot product, the angle between the vectors is given by

$$\theta = \cos^{-1}\left(\frac{\boldsymbol{a} \cdot \boldsymbol{b}}{|\boldsymbol{a}||\boldsymbol{b}|}\right) = \cos^{-1}\left(\frac{1 \times 3 + 2 \times 2 + 3 \times 1}{\sqrt{14}\sqrt{14}}\right) = \cos^{-1}\left(\frac{5}{7}\right).$$

From the definition of the cross product, one has

$$\theta = \sin^{-1}\left(\frac{|\boldsymbol{a} \times \boldsymbol{b}|}{|\boldsymbol{a}||\boldsymbol{b}|}\right),$$

where the magnitude of the cross product is

$$|\boldsymbol{a} \times \boldsymbol{b}| = \left|\begin{pmatrix} 1 \\ 2 \\ 3 \end{pmatrix} \times \begin{pmatrix} 3 \\ 2 \\ 1 \end{pmatrix}\right|$$

$$= \left|\begin{pmatrix} 2 \times 1 - 3 \times 2 \\ 3 \times 3 - 1 \times 1 \\ 1 \times 2 - 2 \times 3 \end{pmatrix}\right| = \sqrt{(-4)^2 + 8^2 + (-4)^2} = 4\sqrt{6}.$$

Thus, one has

$$\theta = \sin^{-1}\left(\frac{4\sqrt{6}}{\sqrt{14}\sqrt{14}}\right) = \sin^{-1}\left(\frac{2\sqrt{6}}{7}\right).$$

One may check numerically that the two derived values of $\theta$ are equal. Alternatively, one may use the triangle of Figure C.1.

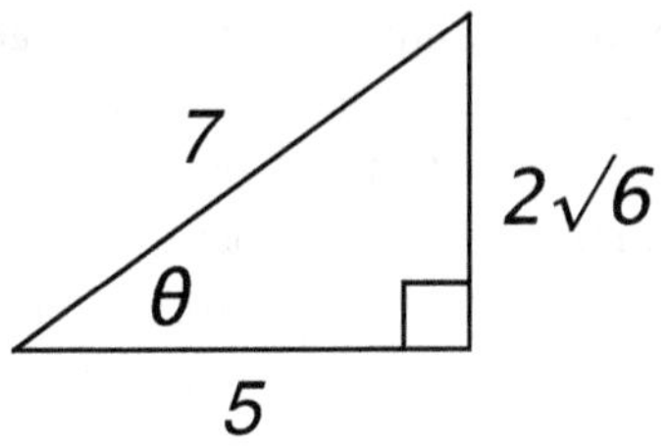


Fig. C.1 Right-angled triangle from which one may prove that $\sin^{-1}(2\sqrt{6}/7) = \cos^{-1}(5/7)$.

(4) In two dimensions two vectors $\boldsymbol{a}$ and $\boldsymbol{b}$ define a plane, and there is no perpendicular direction in which to define a vector cross-product. However, one may still define a scalar that has magnitude $|\boldsymbol{a}||\boldsymbol{b}|\sin\theta$, where $\theta$ is the angle between the vectors. In four or more dimensions, the cross-product becomes ambiguous as there is more than one perpendicular direction for a given pair of vectors. Thus, the cross-product is not used in higher dimensions.

## Chapter 3: Introducing Electricity

(1) The field due to a system of point charges can be obtained by superposing the field due to the individual charges, as stated in Eq. (3.11). In the present case one has

$$\boldsymbol{x} - \boldsymbol{x}_1 = \begin{pmatrix} x \\ 0 \end{pmatrix} - \begin{pmatrix} 0 \\ L \end{pmatrix} = x\hat{\boldsymbol{\imath}} - L\hat{\boldsymbol{\jmath}},$$

$$\boldsymbol{x} - \boldsymbol{x}_2 = \begin{pmatrix} x \\ 0 \end{pmatrix} - \begin{pmatrix} 0 \\ -L \end{pmatrix} = x\hat{\boldsymbol{\imath}} + L\hat{\boldsymbol{\jmath}}.$$

Furthermore, we have

$$|\boldsymbol{x} - \boldsymbol{x}_1| = |\boldsymbol{x} - \boldsymbol{x}_2| = (x^2 + L^2)^{1/2}.$$

Substituting these results into the formula for the electric field gives

$$\boldsymbol{E} = \frac{1}{4\pi\epsilon_0 (x^2 + L^2)^{3/2}}[(Q_1 + Q_2)x\hat{\boldsymbol{\imath}} + L(Q_2 - Q_1)\hat{\boldsymbol{\jmath}}]$$

as required.

(2) If a vector $\boldsymbol{l}$ points from a charge $-q$ to a charge $q$ $(q > 0)$, the dipole moment is given by $\mu = q\boldsymbol{l}$. A dipole in a constant electric field experiences the torque of Eq. (3.14). Provided the distance between the charges is small compared to other distance scales in the problem, we can assume that the electric field is approximately constant over the dipole, so that this result applies. Then, from the result of problem 1, the torque on the dipole is

$$\begin{aligned}\boldsymbol{G} &= \frac{1}{4\pi\epsilon_0(x^2+L^2)^{3/2}}\begin{pmatrix}\mu\\0\\0\end{pmatrix}\times\begin{pmatrix}x(Q_1+Q_2)\\L(Q_2-Q_1)\\0\end{pmatrix}\\ &= \frac{1}{4\pi\epsilon_0(x^2+L^2)^{3/2}}\begin{pmatrix}0\\0\\\mu L(Q_2-Q_1)\end{pmatrix}\\ &= \frac{\mu L(Q_2-Q_1)}{4\pi\epsilon_0(x^2+L^2)^{3/2}}\hat{\boldsymbol{k}}.\end{aligned}$$

If the dipole is not small, then we cannot assume that the electric field is constant across it. This invalidates the use of Eq. (3.14), such that we would have to do a much more careful analysis of how each individual charge in the dipole moves.

(3) Consider a length $dx$ of the wire, at position

$$\boldsymbol{x} = \begin{pmatrix}x\\0\end{pmatrix},$$

and let the position of the point $P$ be

$$\boldsymbol{y} = \begin{pmatrix}0\\y\end{pmatrix}.$$

The field at $P$ due to the segment of wire $dx$ is then given by

$$d\boldsymbol{E} = \frac{dQ}{4\pi\epsilon_0}\frac{(\boldsymbol{y}-\boldsymbol{x})}{|\boldsymbol{y}-\boldsymbol{x}|^3},$$

where $dQ$ is the charge of the segment. Given that the rod is uniformly charged, we may write the latter as

$$dQ = \lambda dx,$$

where $\lambda$ is a constant charge per unit length. From the above vectors one has

$$\boldsymbol{y}-\boldsymbol{x}=\begin{pmatrix}-x\\ y\end{pmatrix}, \quad |\boldsymbol{y}-\boldsymbol{x}|=(x^2+y^2)^{1/2},$$

so that

$$d\boldsymbol{E}=\frac{\lambda}{4\pi\epsilon_0}\frac{dx}{(x^2+y^2)^{3/2}}\begin{pmatrix}-x\\ y\end{pmatrix}.$$

The total electric field is then

$$\boldsymbol{E}=\frac{\lambda}{4\pi\epsilon_0}\int_{-L/2}^{L/2}\frac{dx}{(x^2+y^2)^{3/2}}\begin{pmatrix}-x\\ y\end{pmatrix}.$$

We may integrate each component of this equation separately, and the $x$-component is

$$\begin{aligned}E_x&=\frac{\lambda}{4\pi\epsilon_0}\int_{-L/2}^{L/2}dx\frac{(-x)}{(x^2+y^2)^{3/2}}\\&=\frac{\lambda}{4\pi\epsilon_0}\left[\frac{1}{(x^2+y^2)^{1/2}}\right]_{-L/2}^{L/2}\\&=0.\end{aligned}$$

This is not surprising — by symmetry, the $x$ component of the total field at $P$ must vanish, as it receives equal contributions pointing to the left and to the right!

For the $y$-component, we have

$$\begin{aligned}E_y&=\frac{\lambda}{4\pi\epsilon_0}\int_{-L/2}^{L/2}dx\frac{y}{(x^2+y^2)^{3/2}}\\&=\frac{\lambda}{4\pi\epsilon_0}\left[\frac{x}{y(x^2+y^2)^{1/2}}\right]_{-L/2}^{L/2}\\&=\frac{\lambda}{4\pi\epsilon_0}\frac{L}{y(\frac{L^2}{4}+y^2)^{1/2}}\\&=\frac{\lambda L}{2\pi\epsilon_0}\frac{1}{y(L^2+4y^2)^{1/2}}.\end{aligned}$$

Finally, we may recognise $Q = \lambda L$ as the total charge of the rod. The $x$ and $y$-components may then be summarised in the single equation

$$\boldsymbol{E} = \frac{Q}{2\pi\epsilon_0 y(L^2 + 4y^2)^{1/2}}\hat{\jmath}$$

as given.

(4) Gauss' law states that

$$\Phi_E = \frac{Q}{\epsilon_0},$$

where $\Phi_E$ is the total electric flux through a closed surface $S$, and $Q$ the total charge enclosed by $S$. In this case there is an electron inside the cube, so that the total charge enclosed is $Q = -1.6\times 10^{-19}$ C. If the electron is at the centre of the cube, then the flux through each surface of the cube must be equal, by symmetry. A cube has six faces, so that the flux through each face is

$$\frac{1}{6}\Phi_E = \frac{1}{6}\frac{1.6\times 10^{-19}}{8.85\times 10^{-12}} = -1.81\times 10^{-8}\,\mathrm{NC^{-1}\,m^2}.$$

This is negative, as the charge is negative so that the electric field points towards the charge, and so represents a flux *into* the surface.

(5) Consider the sphere containing the charge to be centred on the origin. The system has spherical symmetry, so that we may apply Gauss' law to a spherical surface whose radius $r > R$ to find the field outside the sphere of radius $R$. By the symmetry, the electric field will be constant in magnitude on the larger sphere, and the flux will thus be

$$\Phi_E = \int\int d\boldsymbol{S}\cdot\boldsymbol{E} = |\boldsymbol{E}|\int\int dS = 4\pi r^2|\boldsymbol{E}|.$$

Gauss' Law then gives

$$|\boldsymbol{E}| = \frac{Q}{4\pi\epsilon_0 r^2},$$

where $Q$ is the total charge enclosed within the sphere of radius $r$, hence also by the sphere of radius $R$. The direction of the field will be radially outwards, again by the symmetry of the problem. Thus, the field is the same as one expects for a point charge located at the origin. For a non-spherically symmetric charge distribution, it is no longer straightforward to carry out the surface integral, as the electric field will no longer be constant on a spherical surface.

(6) Given the spherical symmetry of the problem, we may again apply Gauss' Law using a spherical surface. Inside the sphere, we can choose a sphere of radius $r < R$ to be our Gaussian surface. As in problem 1, the flux will be

$$\Phi_E = 4\pi r^2 |\boldsymbol{E}|,$$

and the field will point radially outwards. To find the total charge enclosed by the surface, consider a thin spherical shell of radius

$$0 < r' \leq R.$$

The volume of this shell is

$$dV = 4\pi r'^2 dr',$$

so that the charge contained in the shell is

$$\rho(r')dV = 4\pi r'^2 \rho(r')dr'.$$

The total charge enclosed up to radius $r$ can then be found by adding together the contributions from each shell, yielding

$$\begin{aligned} Q(r) &= \int_0^r 4\pi r'^2 \rho(r')dr' \\ &= 4\pi \int_0^r \lambda r'^4 \\ &= 4\pi\lambda \left[ \frac{r'^5}{5} \right]_0^r \\ &= \frac{4\pi\lambda r^5}{5}. \end{aligned}$$

Gauss' Law then gives

$$4\pi r^2 |\boldsymbol{E}| = \frac{4\pi\lambda r^5}{5},$$

and thus

$$|\boldsymbol{E}| = \frac{\lambda r^3}{5\epsilon_0}, \quad 0 \leq r < R.$$

For $r > R$, Gauss' Law becomes

$$4\pi r^2 |\boldsymbol{E}| = \frac{4\pi\lambda R^5}{5\epsilon_0}$$

i.e. the total charge enclosed gets capped by the fact that there is no charge for $r > R$. One thus has

$$|\boldsymbol{E}| = \frac{\lambda R^5}{5\epsilon_0 r^2},$$

with the direction radially outward as before.

(7) The electric field is given by Eq. (3.35), which gives

$$\begin{aligned} E_x &= -\alpha\left[\frac{1}{x^2+y^2+L^2} - \frac{2x(x+y-L)}{(x^2+y^2+L^2)^2}\right] \\ &= -\alpha\frac{L^2+2Lx-x^2-2xy+y^2}{(L^2+x^2+y^2)^2}; \\ E_y &= -\alpha\left[\frac{1}{x^2+y^2+L^2} - \frac{2y(x+y-L)}{(x^2+y^2+L^2)^2}\right] \\ &= -\alpha\frac{L^2+2Ly+x^2-2xy-y^2}{(L^2+x^2+y^2)^2}; \\ E_z &= 0. \end{aligned}$$

## Chapter 4: A First Look at Circuits

(1) The electric field between the spheres is similar to that of the previous question, but where $Q$ is replaced by $-Q$. To see this, note that one would find the field between the spheres using Gauss' Law, which only cares about the charged enclosed inside a spherical surface drawn around the inner sphere of the capacitor. The charge on the latter is $-Q$.

We can then find the potential difference from the inner to the outer sphere by recycling an intermediate result from the previous question, and sending $Q \to -Q$:

$$\begin{aligned} \Delta V &= \frac{Q}{4\pi\epsilon_0}\left[-\frac{1}{r}\right]_{R_1}^{R_2} \\ &= \frac{Q}{4\pi\epsilon_0}\left(\frac{1}{R_1} - \frac{1}{R_2}\right) \\ &= \frac{Q}{4\pi\epsilon_0}\frac{R_2 - R_1}{R_1 R_2}. \end{aligned}$$

From $Q = C\Delta V$, one then has

$$C = \frac{Q}{\Delta V} = \frac{4\pi\epsilon_0 R_1 R_2}{R_2 - R_1}.$$

(3) (i) From Eq. (4.11), we find

$$E = \frac{(1.68 \times 10^{-8}) \times 6}{1 \times 10^{-6}} = 0.1\,\mathrm{Vm}^{-1}.$$

(ii) From Eq. (4.7), the magnitude of the drift velocity is given by

$$|\boldsymbol{v}_d| = \frac{J}{ne} = \frac{I}{Ane}.$$

Plugging in the numbers gives

$$|\boldsymbol{v}_d| = \frac{6}{(1 \times 10^{-6}) \times (8.5 \times 10^{28}) \times (1.6 \times 10^{-19})}$$
$$\simeq 4.4 \times 10^{-4}\,\mathrm{ms}^{-1}.$$

This is extremely small compared to a typical human walking speed of between 1 and 2 $\mathrm{ms}^{-1}$! We see therefore that the electrons do not have to travel very fast to carry a current.

## Chapter 5: Introducing Magnetism

(1) An electron undergoing circular motion with speed $v$ and radius $R$ has a constant magnetic field of magnitude $B$ perpendicular to its motion. The Lorentz force law gives the magnitude of the force as

$$F = evB = m\frac{v^2}{R},$$

where on the right we have equated the force to the general expression for the centripetal force in uniform circular motion, where $m$ is the mass of the particle. Rearranging gives an expression for the radius:

$$R = \frac{mv}{eB} = \frac{(9.11 \times 10^{-31}) \times (0.5 \times 3 \times 10^8)}{(1.6 \times 10^{-19}) \times 1} = 8.5 \times 10^{-4}\,\mathrm{m}.$$

The speed is related to the angular frequency $\omega$ by

$$v = \omega R,$$

and thus one has

$$\omega = \frac{eB}{m}.$$

Then the period is given by

$$T = \frac{2\pi}{\omega} = \frac{2\pi m}{eB}.$$

Plugging in the numbers gives $T = 3.6 \times 10^{-11}$s.

(2) The magnetic field generated at charge $q_2$ due to charge $q_1$ is

$$\boldsymbol{B}_{21} = \frac{\mu_0 q_1}{4\pi} \frac{\boldsymbol{v}_1 \times (\boldsymbol{x}_2 - \boldsymbol{x}_1)}{|\boldsymbol{x}_2 - \boldsymbol{x}_1|^3}.$$

Then the Lorentz force law gives the force on $q_2$ due to $q_1$ as

$$\boldsymbol{F}_{21} = q_2 \boldsymbol{v}_2 \times \boldsymbol{B}_{21} = \frac{\mu_0 q_1 q_2}{4\pi} \frac{\boldsymbol{v}_2 \times [\boldsymbol{v}_1 \times (\boldsymbol{x}_2 - \boldsymbol{x}_1)]}{|\boldsymbol{x}_2 - \boldsymbol{x}_1|^3}$$

as required.

(3) Using the result of exercise (2), the force on charge $q_1$ due to $q_2$ is

$$\boldsymbol{F}_{12} = \frac{\mu_0 q_1 q_2}{4\pi} \frac{\boldsymbol{v}_1 \times [\boldsymbol{v}_2 \times (\boldsymbol{x}_1 - \boldsymbol{x}_2)]}{|\boldsymbol{x}_2 - \boldsymbol{x}_1|^3}$$

$$= -\frac{\mu_0 q_1 q_2}{4\pi} \frac{\boldsymbol{v}_1 \times [\boldsymbol{v}_2 \times (\boldsymbol{x}_2 - \boldsymbol{x}_1)]}{|\boldsymbol{x}_2 - \boldsymbol{x}_1|^3},$$

and thus

$$\boldsymbol{F}_{12} \neq -\boldsymbol{F}_{21},$$

due to the fact that

$$\boldsymbol{v}_2 \times [\boldsymbol{v}_1 \times (\boldsymbol{x}_2 - \boldsymbol{x}_1)] \neq \boldsymbol{v}_1 \times [\boldsymbol{v}_2 \times (\boldsymbol{x}_2 - \boldsymbol{x}_1)]$$

in general. To analyse thus further, we can use the vector identity

$$\boldsymbol{a} \times (\boldsymbol{b} \times \boldsymbol{c}) = (\boldsymbol{a} \cdot \boldsymbol{c})\boldsymbol{b} - (\boldsymbol{a} \cdot \boldsymbol{b})\boldsymbol{c}$$

to write

$$\boldsymbol{v}_2 \times [\boldsymbol{v}_1 \times (\boldsymbol{x}_2 - \boldsymbol{x}_1)] = \boldsymbol{v}_2 \cdot (\boldsymbol{x}_2 - \boldsymbol{x}_1)\boldsymbol{v}_1 - \boldsymbol{v}_1 \cdot \boldsymbol{v}_2(\boldsymbol{x}_2 - \boldsymbol{x}_1),$$

$$\boldsymbol{v}_1 \times [\boldsymbol{v}_2 \times (\boldsymbol{x}_2 - \boldsymbol{x}_1)] = \boldsymbol{v}_1 \cdot (\boldsymbol{x}_2 - \boldsymbol{x}_1)\boldsymbol{v}_2 - \boldsymbol{v}_1 \cdot \boldsymbol{v}_2(\boldsymbol{x}_2 - \boldsymbol{x}_1).$$

There are then a number of special cases where Newton's third law indeed applies:

(i) $\boldsymbol{v}_1 = \boldsymbol{v}_2 = \boldsymbol{0}$. This is a trivial case, as the magnetic force is zero on both particles!

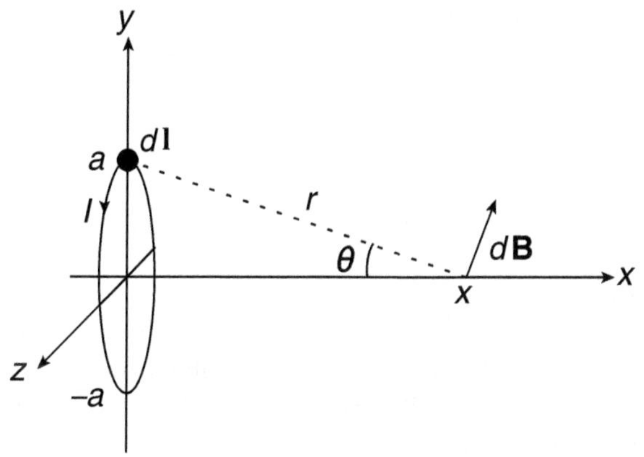


Fig. C.2 A current-carrying ring, with a segment $\boldsymbol{dl}$ pointing in the direction of the current.

(ii) $\boldsymbol{v}_2 \cdot (\boldsymbol{x}_2 - \boldsymbol{x}_1) = \boldsymbol{v}_1 \cdot (\boldsymbol{x}_2 - \boldsymbol{x}_1) = 0$, i.e. the velocity of each particle is perpendicular to the displacement between the particles. The case of two current-carrying wires is an example of this.

(iii) $\boldsymbol{v}_1 = \boldsymbol{v}_2$, i.e. both velocities are equal.

A violation of Newton's third law implies that there might be a problem with momentum conservation, given that the force on a given particle is equal to the rate of change of its momentum: if the forces on two particles are not equal and opposite, then they can gain momentum such that the total momentum of both particles is not constant. However, there is also momentum carried by the electric and magnetic fields. When this is taken properly into account, the conservation of total momentum is restored.

(4) (a) The contribution to the magnetic field at a general point along the $x$-axis is shown in Figure C.2. That is, $\boldsymbol{dl}$ at any segment on the ring is perpendicular to the displacement $\boldsymbol{r}$ from the segment to the point $\boldsymbol{x}$. The right-hand rule then tells that, upon rotating the segment into the displacement vector, the resulting vector is tilted from the vertical by the angle $\theta$ shown in Figure C.2. The magnitude of the field is given by the Biot–Savart law of Eq. (5.9), and the $x$-component can be obtained, using trigonometry, as

$$dB_x = \frac{\mu_0 I}{4\pi} \frac{|\boldsymbol{dl} \times \boldsymbol{r}| \sin\theta}{r^3} = \frac{\mu_0 I}{4\pi} \frac{dl}{r^2} \sin\theta,$$

where we have used the fact that $\boldsymbol{dl} \perp \boldsymbol{r}$, so that $|\boldsymbol{dl} \times \boldsymbol{r}| = rdl$.

Also from the figure, we find

$$\sin\theta = \frac{a}{r}, \quad r = \sqrt{a^2 + x^2},$$

so that

$$dB_x = \frac{\mu_0 I a}{4\pi}\frac{dl}{(x^2 + a^2)^{3/2}}$$

as required.

(b) The total field is obtained by integrating over all segments of the ring. By symmetry, the $y$ and $z$ components of the field must cancel out upon combining all contributions. Thus, one is left with a field purely in the $x$-direction:

$$\boldsymbol{B} = \hat{\boldsymbol{\imath}}\int_0^L dl \frac{\mu_0 I a}{4\pi(x^2 + a^2)^{3/2}} = \hat{\boldsymbol{\imath}}\frac{\mu_0 I a}{4\pi(x^2 + a^2)^{3/2}}$$
$$\times \int_0^L dl = \hat{\boldsymbol{\imath}}\frac{\mu_0 I a L}{4\pi(x^2 + a^2)^{3/2}},$$

where $L$ is the total length of the ring. This is given by

$$L = 2\pi a,$$

so that one finally has

$$\boldsymbol{B} = \frac{\mu_0 I a^2}{2(x^2 + a^2)^{3/2}}\hat{\boldsymbol{\imath}}.$$

(5) A current loop has a magnetic dipole moment

$$\boldsymbol{\mu}_B = I\boldsymbol{S},$$

where $\boldsymbol{S}$ is the vector area of the minimal surface spanned by the loop, and $I$ the current. The torque on the loop is given by

$$\boldsymbol{G} = \boldsymbol{\mu}_B \times \boldsymbol{B},$$

so that

$$|\boldsymbol{G}| = |\boldsymbol{\mu}_B||\boldsymbol{B}|\sin\theta,$$

with $\theta$ the angle between the normal to the loop, and the magnetic field. In the present case we have

$$|\boldsymbol{S}| = \pi R^2,$$

so that

$$|\boldsymbol{G}| = \pi R^2 IB \sin\theta = \pi \times 0.02^2 \times 1 \times 1 \times \sin(\pi/4) = 8.9 \times 10^{-4}\,\text{Nm}.$$

(6) Using the right-hand rule in part (b) of the figure, the magnetic field will be pointing in the $-y$-direction (to the left) above the sheet, and in the $+y$ direction (to the right) below. Using the given contour, the contribution to the line integral will be zero on the two vertical segments, as the field is perpendicular to the contour there. On the upper and lower segments the field is parallel to the contour. It will also be constant in magnitude, as by symmetry it must only depend on the perpendicular distance from the sheet. The line integral thus gives

$$\oint \boldsymbol{B} \cdot d\boldsymbol{l} = 2BL.$$

By Ampère's law, this must be equal to $\mu_0$ times the total current enclosed by the contour, which is $IL$, given that $I$ is the current per unit length along the $y$-direction. This gives

$$B = \frac{\mu_0 I}{2}$$

as required.

## Chapter 6: A Second Look at Circuits

(1) The time constant of a DC R-C circuit is given by

$$\tau = RC,$$

and the capacitor has mostly discharged for $t > \tau$. Plugging in the numbers gives $4\,\mu$s.

(2) (a) The potential difference across each inductor is given by

$$V_i = L_i \frac{dI}{dt},$$

where, by conservation of charge, the *same* current must flow through both inductors. The overall potential difference across both inductors is then

$$V = V_1 + V_2 = (L_1 + L_2)\frac{dI}{dt}.$$

The system thus behaves like a single inductor, with inductance

$$L = L_1 + L_2.$$

(b) Let each inductor have length $l_i$, and number of turns $N_i$. The number of turns per unit length,

$$n = \frac{N_i}{l_i},$$

is the same for each solenoid. Thus, we may rewrite the inductance of solenoid $i$ as

$$L_i = \mu_0 \pi R^2 n^2 l_i.$$

Adding together the inductances of both solenoids gives

$$L_1 + L_2 = \mu_0 \pi R^2 n^2 (l_1 + l_2).$$

Thus, the compound system indeed looks like a single inductor, with the effective inductance associated with a length equal to the sums of the lengths of the individual solenoids, as it should be.

(3) The case of impedances in series implies that the complex impedance of two capacitors in series is

$$Z = Z_1 + Z_2 = -i\left(\frac{1}{\omega C_1} + \frac{1}{\omega C_2}\right),$$

where we have used the result of Eq. (6.28). If we want to view this as the impedance of a single capacitor with capacitance $C$, we should write it as

$$Z = -\frac{i}{\omega C}, \tag{C.1}$$

which implies

$$\frac{1}{C} = \frac{1}{C_1} + \frac{1}{C_2},$$

thus reproducing Eq. (4.20). Similarly, the rule for impedances in parallel implies that the overall impedance satisfies

$$\frac{1}{Z} = \frac{1}{Z_1} + \frac{1}{Z_2} = i\omega C_1 + i\omega C_2.$$

If we are to view this as a single impedance for a capacitor $C$, Eq. (C.1) now implies

$$C = C_1 + C_2,$$

which reproduces Eq. (4.21).

## Chapter 7: Maxwell's Equations

(1) We may use the form of the divergence and curl in Cartesian coordinates from Eqs. (7.11) and (7.13), to get

$$\begin{aligned}\nabla \cdot (\nabla \times \boldsymbol{F}) &= \frac{\partial}{\partial x}\left(\frac{\partial F_z}{\partial y} - \frac{\partial F_y}{\partial z}\right) + \frac{\partial}{\partial y}\left(\frac{\partial F_x}{\partial z} - \frac{\partial F_z}{\partial x}\right) \\ &\quad + \frac{\partial}{\partial z}\left(\frac{\partial F_y}{\partial x} - \frac{\partial F_x}{\partial y}\right) \\ &= \left(\frac{\partial^2 F_z}{\partial x \partial y} - \frac{\partial^2 F_z}{\partial y \partial x}\right) + \left(\frac{\partial^2 F_x}{\partial y \partial z} - \frac{\partial^2 F_x}{\partial z \partial y}\right) \\ &\quad + \left(\frac{\partial^2 F_y}{\partial z \partial x} - \frac{\partial^2 F_y}{\partial x \partial z}\right) \\ &= 0,\end{aligned}$$

where we have used the fact that the order in which partial derivatives are applied is unimportant.

(2) The vacuum Maxwell equations are given by Eqs. (7.22)–(7.25). Upon making the suggested transformations, the first two equations become

$$-c(\nabla \cdot \boldsymbol{B}) = \frac{1}{c}(\nabla \cdot \boldsymbol{E}) = 0 \;\Rightarrow\; \nabla \cdot \boldsymbol{E} = \nabla \cdot \boldsymbol{B} = 0,$$

as before. For the third equation one finds

$$-c\nabla \times \boldsymbol{B} = -\frac{1}{c}\frac{\partial \boldsymbol{E}}{\partial t} \;\Rightarrow\; \nabla \times \boldsymbol{B} = \frac{1}{c^2}\frac{\partial \boldsymbol{E}}{\partial t},$$

which becomes Eq. (7.25) upon using Eq. (7.27). Finally, for the fourth equation one finds

$$\frac{1}{c}\nabla\times\boldsymbol{E}=-c\mu_0\epsilon_0\frac{\partial\boldsymbol{B}}{\partial t}\quad\Rightarrow\quad\nabla\times\boldsymbol{E}=-c^2\mu_0\epsilon_0\frac{\partial\boldsymbol{B}}{\partial t},$$

which becomes Eq. (7.24) upon using Eq. (7.27).

We have thus shown that the vacuum Maxwell equations remain the same under the duality transformation. In the non-vacuum case, the symmetry is broken by the presence of charges and currents: two of the equations have these in, but the other two do not.

(3) A general linearly polarised light wave travelling in the $z$-direction will have an electric field of the form

$$\boldsymbol{E}=\mathrm{Re}[(A_1\hat{\mathbf{e}}_1+A_2\hat{\mathbf{e}}_2)\,e^{i(\omega t-kz)}],$$

so that the action of the quarter-wave plate is to turn this into

$$\boldsymbol{E}=\mathrm{Re}[(A_1\hat{\mathbf{e}}_1+e^{i\pi/2}A_2\hat{\mathbf{e}}_2)e^{i(\omega t-kz)}].$$

A relative phase difference of $\pi/2$ will lead to elliptically polarised light in general, unless the amplitudes in the $x$ and $y$ directions are the same. One can guarantee this by arranging that the incoming linearly polarised light is at 45° to the $x$-axis.

## Chapter 8: Relativity and Maxwell's Equations

(1) Let us consider measuring the length of the (moving) object along the $x$-direction in $S$. To make this formal, let us take the endpoints of the object to be at $x_1$ and $x_2$, considered at times $t_1$ and $t_2$ in general. These correspond to positions and times $(x_1', t_1')$ and $(x_2', t_2')$ in $S'$, as given by Eq. (5.2). Applying the Lorentz transformations to both sets of points and taking the difference, we obtain

$$\begin{aligned}
c\Delta t' &= \gamma\left(c\Delta t-\frac{v\Delta x}{c}\right);\\
\Delta x' &= \gamma\,(\Delta x - v\Delta t)\,;\\
\Delta y' &= \Delta y;\\
\Delta z' &= \Delta z,
\end{aligned}$$

where $\Delta x = x_2 - x_1$, etc. Now note that a length measurement in $S$ consists of measuring the distance $\Delta x = L$ between both ends of the object at the *same time*, such that $\Delta t = 0$. We thus immediately find

$$\Delta x' = \gamma \Delta x = \gamma L.$$

The left-hand side contains the difference in position of the object considered at *different times* $t_1'$ and $t_2'$ in the frame $S'$. However, the object is at rest in $S'$, and thus it does not matter at what time we consider either of the endpoints of the object. This allows us to identify $\Delta x' = L_0$, such that

$$L_0 = \gamma L \;\Rightarrow\; L = \frac{L_0}{\gamma}.$$

For the time dilation result, imagine we measure a time interval in $S'$ at the same position along the $x'$-direction, so that $\Delta x' = 0$. From the above formulae we then obtain

$$\Delta x = v\Delta t \quad \Rightarrow \quad c\Delta t' = \gamma\left(c\Delta t - \frac{v^2 \Delta t}{c}\right).$$

Identifying $\Delta t' = \tau_0$, $\Delta t = \tau$, we thus find

$$\tau_0 = \gamma\left(1 - \frac{v^2}{c^2}\right)\tau = \frac{\tau}{\gamma},$$

where we have used Eq. (8.3). Rearranging gives $\tau = \gamma\tau_0$ as required.

(2) From Eqs. (3.34) and (7.13), one finds

$$\begin{aligned}
\nabla \times (\nabla f) &= \begin{pmatrix} \partial/\partial x \\ \partial/\partial y \\ \partial/\partial z \end{pmatrix} \times \begin{pmatrix} \partial f/\partial x \\ \partial f/\partial y \\ \partial f/\partial z \end{pmatrix} \\
&= \begin{pmatrix} \partial^2 f/\partial y \partial z - \partial^2 f/\partial z \partial y \\ \partial^2 f/\partial z \partial x - \partial^2 f/\partial x \partial z \\ \partial^2 f/\partial x \partial y - \partial^2 f/\partial y \partial x \end{pmatrix} \\
&= \begin{pmatrix} 0 \\ 0 \\ 0 \end{pmatrix},
\end{aligned}$$

where in the final line we have used the fact that the order of partial derivatives is unimportant.

(3) One has

$$F^{\mu}{}_{\nu} = \eta^{\mu\alpha} F_{\alpha\nu},$$

where $\eta^{\mu\alpha}$ is the inverse metric tensor of Eq. (8.42). In order to interpret any index equation as a matrix equation, we must make sure that contracted indices line up so that they are next to each other (sometimes referred to as the *domino rule* of matrix multiplication). In fact, this is already the case above, so that we may straightforwardly identify that the components of $F^{\mu}{}_{\nu}$ are given by

$$F^{\mu}{}_{\nu} = \begin{pmatrix} 1 & 0 & 0 & 0 \\ 0 & -1 & 0 & 0 \\ 0 & 0 & -1 & 0 \\ 0 & 0 & 0 & -1 \end{pmatrix} \begin{pmatrix} 0 & E_x/c & E_y/c & E_z/c \\ -E_x/c & 0 & -B_z & B_y \\ -E_y/c & B_z & 0 & -B_x \\ -E_z/c & -B_y & B_x & 0 \end{pmatrix}$$

$$= \begin{pmatrix} 0 & E_x/c & E_y/c & E_z/c \\ E_x/c & 0 & B_z & -B_y \\ E_y/c & -B_z & 0 & B_x \\ E_z/c & B_y & -B_x & 0 \end{pmatrix}.$$

Next, we have

$$F^{\mu\nu} = F^{\mu}{}_{\nu} \eta^{\alpha\nu}.$$

Now the contracted indices are not next to each other. However, we may use the fact that the transpose of a matrix may be expressed in index notation as

$$(M^{\mathrm{T}})^{\nu\alpha} = M^{\alpha\nu}.$$

In other words, we may switch round the indices of any matrix, provided we replace the matrix by its transpose. However, the inverse metric of Eq. (8.42) is symmetric, so that it is equal to its transpose. We can thus rewrite the above as

$$F^{\mu\nu} = F^{\mu}{}_{\nu} \eta^{\nu\alpha},$$

which corresponds to multiplying the matrix components of $F^{\mu}{}_{\nu}$ on the right with the matrix of Eq. (8.42). One finds

$$F^{\mu\nu} = \begin{pmatrix} 0 & E_x/c & E_y/c & E_z/c \\ E_x/c & 0 & B_z & -B_y \\ E_y/c & -B_z & 0 & B_x \\ E_z/c & B_y & -B_x & 0 \end{pmatrix} \begin{pmatrix} 1 & 0 & 0 & 0 \\ 0 & -1 & 0 & 0 \\ 0 & 0 & -1 & 0 \\ 0 & 0 & 0 & -1 \end{pmatrix}$$

$$= \begin{pmatrix} 0 & E_x/c & E_y/c & E_z/c \\ E_x/c & 0 & -B_z & B_y \\ E_y/c & B_z & 0 & -B_x \\ E_z/c & -B_y & B_x & 0 \end{pmatrix}.$$

(4) Let us first invert Eqs. (8.86) and (8.87) to get

$$E_x = E'_x, \quad E_y = \gamma(E'_y + vB'_z), \quad E_z = \gamma(E'_z - vB'_y)$$

and

$$B_x = B'_x, \quad B_y = \gamma\left(B'_y - \frac{v}{c^2}E'_z\right), \quad B_z = \gamma\left(B'_z + \frac{v}{c^2}E'_y\right).$$

These equations can be straightforwardly obtained by interchanging the primed and unprimed fields in Eqs. (8.86) and (8.87), and sending $v \to -v$. Next, note that if the point charge is at rest in $S'$, there is no magnetic field, so that $B'_x = B'_y = B'_z = 0$, and

$$B_x = 0, \quad B_y = -\frac{\gamma v}{c^2}E'_z, \quad B_z = \frac{\gamma v}{c^2}E'_y.$$

For a positive point charge, the electric field in $S'$ will point radially outwards from the origin. Then the magnetic field in $S$ points azimuthally around the $x$-direction, using a right-hand rule. This makes sense: in the frame $S$, the moving point charge constitutes a current, and the direction of the magnetic field found here indeed matches the right-hand rule.

## Chapter 9: Maxwell's Equations from Symmetry

(1) From Eqs. (8.63) and (9.7) we have

$$A_\mu + \delta A_\mu = \left(\frac{V}{c}, -\boldsymbol{A}\right) - \frac{1}{e}\left(\frac{1}{c}\frac{\partial \alpha}{\partial t}, \nabla\alpha\right)$$
$$= \left(\frac{V}{c} - \frac{1}{e}\frac{1}{c}\frac{\partial\alpha}{\partial t}, -\boldsymbol{A} - \frac{1}{e}\nabla\alpha\right),$$

where we have used Eq. (8.64). Rearranging then gives the required transformations, and we may note that these forms are correct even if we are not in natural units (where $c = 1$).

(2) (a) In a vacuum, Eq. (8.78) reduces to

$$\partial^\mu F_{\mu\nu} = 0.$$

Using Eq. (8.67) for the field strength tensor, this can be written as

$$\partial^\mu(\partial_\mu A_\nu - \partial_\nu A_\mu) = \partial_\mu \partial^\mu A_\nu - \partial_\nu(\partial^\mu A_\mu) = 0,$$

where we have interchanged the order of partial derivatives in places. As noted in Eq. (8.35), we can always swap which index is upstairs and which is downstairs in any particular contraction of indices. In particular, we may write

$$\partial^\mu A_\mu = \partial_\mu A^\mu = 0.$$

Thus, we end up with

$$\partial_\mu \partial^\mu A_\nu = 0.$$

(b) In natural units one has

$$\partial_\mu \partial^\mu \equiv \frac{\partial^2}{\partial t^2} - \nabla^2.$$

Thus, the result of part (a) becomes

$$\nabla^2 A_\nu = \frac{\partial^2 A_\nu}{\partial t^2}.$$

This is the wave equation, whose solutions are the same electromagnetic waves that we discovered in Chapter 7.

## Chapter 10: The Double Copy

(1) One has

$$g_{\mu\nu} = \eta_{\mu\nu} + \kappa h_{\mu\nu}, \quad h_{\mu\nu} = \phi k_\mu k_\nu,$$

where $\eta_{\mu\nu}$ is the metric of flat space with $\eta_{00} = 1$, and $\phi$ and $k_\mu$ are given by Eq. (10.17). We then have

$$g_{00} = 1 + \frac{\kappa^2}{2}\frac{M}{4\pi r}k_0 k_0 = 1 + \frac{2G_N M}{r},$$

where we have used the fact that $\kappa^2 = 16\pi G_N$. This indeed gives the right form for $V_G$.

(2) The first replacement takes the coupling constant of gravity, and replaces it with its counterpart in a gauge theory. This is exactly the same as in the double copy for scattering amplitudes. The second replacement takes a kinematic quantity in gravity (mass), and replaces it with an amount of charge in the gauge theory. This is very similar in spirit to the double copy for scattering amplitudes, which replaces kinematic information in the gravity theory with (colour) charge information in the gauge theory.

(3) From the given forms of $\alpha$ and the operator $\partial_\mu$, we find

$$\partial_\mu \alpha = \left(0, \frac{Qe^2}{4\pi r}, 0, 0\right),$$

where we have used the fact that $\alpha$ does not depend explicitly on $t$, $\theta$ or $\phi$. The only non-zero component is then simply the ordinary derivative of $\alpha$ with respect to $r$. The gauge transformation of Eq. (9.7) applied to Eq. (10.18) then yields

$$\begin{aligned} A_\mu \to A_\mu - \frac{1}{e}\partial_\mu \alpha &= \left(\frac{Qe}{4\pi r}, \frac{Qe}{4\pi r}, 0, 0\right) - \left(0, \frac{Qe}{4\pi r}, 0, 0\right) \\ &= \left(\frac{Qe}{4\pi r}, 0, 0, 0\right) \end{aligned}$$

as required.

(4) At large distances, the finite size and structure of the disk of charge will not be visible. Thus, the field will look like that of a point charge: static, and pointing radially outwards from the origin.

# Index

## Essential Textbooks in Physics

*(Continued from page ii)*

*Application-Driven Quantum and Statistical Physics: A Short Course for Future Scientists and Engineers*
*Volume 1: Foundations*
by Jean-Michel Gillet

*A Guide to Mathematical Methods for Physicists: With Problems and Solutions*
by Michela Petrini, Gianfranco Pradisi and Alberto Zaffaroni

*Introduction to General Relativity and Cosmology*
by Christian G. Böhmer

*Newtonian Mechanics for Undergraduates*
by Vijay Tymms